Advances in Soil Science

Advances in Soil Science

B.A. Stewart, Editor

Advances in Soil Science

Volume 9

Edited by B.A. Stewart

With Contributions by
P.H.T. Beckett, J. Bouma, M.P.W. Farina,
M.V. Fey, W.P. Miller, M.A. Pavan, I. Shainberg,
M.E. Sumner, K.G. Tiller

With 66 Illustrations

Springer-Verlag
New York Berlin Heidelberg
London Paris Tokyo

B.A. Stewart
USDA Conservation & Production Research Laboratory
Bushland, Texas 79012, USA

ISSN: 0176-9340

Printed on acid-free paper.

Softcover reprint of the hardcover 1st edition 1989

Typeset by Publishers Service, Bozeman, Montana.

9 8 7 6 5 4 3 2 1

ISBN-13: 978-1-4612-8144-3 e-ISBN-13: 978-1-4612-3532-3
DOI: 10.1007/978-1-4612-3532-3

Preface

Soil is formed from the physical and chemical weathering of rocks—processes described historically because they involve eons of time—by glaciation, and by wind and water transport of soil materials, later deposited in deltas and loessial planes. Soil undergoes further transformations over time and provides a habitat for biological life and a base for the development of civilizations. Soil is dynamic—always changing as a result of the forces of nature and particularly by the influences of man. Soils have been studied as long as history has been documented. W.H. Gardner told of writings on clay tablets, dating about 1700 BC, in his review, "Early Soil Physics into the Mid-20th Century," published in Volume 4 of this series. Those writings gave specific instructions on cultivating the soil and seeding crops. Numerous references to soil are found in historical writings, such as Aristotle (384–322 BC), Theophrastus (372–286 BC), Cato the Elder (234–149 BC), and Varro (116–27 BC). Some of the earliest historical references to soil 3000 or more years ago have to do with erosional forces of wind and water.

The study of soils today has taken on increased importance because a rapidly expanding population is placing demands on soil that has never before been experienced. Soil scientists have professionally divided themselves into separate disciplines—physics, chemistry, microbiology, mineralogy, genesis, and the like. Studies range from very basic to very applied, and to literally every corner of the earth, and of the moon as well.

This series, *Advances in Soil Science*, was established to provide a forum for leading scientists to analyze and summarize the available scientific information on a subject, assessing its importance and identifying additional research needs. It is not the purpose of the series to report new research findings because there are many excellent scientific journals for that need. Communications in scientific journals, however, are generally restricted to short and technical presentations. Therefore, *Advances in Soil Science* fills a gap between the scientific journals and the comprehensive reference books in which scientists can delve in depth on a particular subject relating to soil science. Contributors are asked in particular to develop and identify principles that have practical applications to both developing and developed agricultures.

Advances in Soil Science was formulated to be international in scope and to cover all subjects relating to soil science. This volume contains contributions from scientists from Great Britain, The Netherlands, South Africa, Brazil, Israel, Australia, and the United States. These contributions present valuable information on a diversity of topics and serve as an excellent source of references. Although we consider our audience to be primarily scientists and students of soil science, the series provides technical information for anyone interested in our natural resources and man's influence on these resources.

The ultimate aim of the series is to stimulate action: action to determine where there are arable soils, action to develop technology for more efficient crop production on these soils, action to reduce the risk of degrading these soil resources, and action to determine on which soils our research efforts should be concentrated. Research in the future will focus on systems that are resource efficient and environmentally sound. The need to optimize crop production while conserving the resource base has never been greater.

The quick acceptance of *Advances in Soil Science* by both authors and readers has been very gratifying and confirms our perception that a need did exist for a medium to publish soil science reviews. I want to thank the authors for their excellent contributions and cooperation. I also want to thank members of the Editorial Board for their assistance in selecting such competent authors, and the Springer-Verlag staff for their kind assistance. Lastly, and most importantly, I want to thank the readers for their acceptance and use of *Advances in Soil Science*.

B.A. Stewart

Contents

Preface v
Contributors ix

Use of Gypsum on Soils: A Review 1
I. Shainberg, M.E. Sumner, W.P. Miller, M.P.W. Farina, M.A. Pavan, and M.V. Fey
I. Introduction 2
II. Use of Gypsum on Dispersive Soils 6
III. Gypsum As an Ameliorant for Acid Subsoils 54
IV. Environmental Concerns in Phosphogypsum Use 97
V. General Conclusions 100
References 101

Heavy Metals in Soils and Their Environmental Significance 113
K.G. Tiller
I. Introduction 113
II. Sources of Pollution 116
III. Mobility of Metals in Soils and Landscapes 126
IV. Ecological Consequences of Heavy Metal Pollution of Soils 128
V. Prediction of the Impact of Heavy Metal Pollution of Soils 130
VI. Guidelines for Regulatory Control of Pollution of Agricultural Land 133
VII. Final Comment 137
References 137

The Use of Extractants in Studies on Trace Metals in Soils, Sewage Sludges, and Sludge-Treated Soils 143
P.H.T. Beckett
I. Introduction 144
II. Forms of Combination of Trace Metals in Sewage Sludge, Sludged Soils, or Soils 146

III. Survey of Extractants 149
IV. Discussion 168
References 171

Using Soil Survey Data for Quantitative Land Evaluation 177
J. Bouma
I. Introduction 177
II. Soil Survey Interpretation and Land Evaluation 179
III. Modern Land Use Questions 185
IV. Innovative Uses of Soil Survey Data 192
V. Future Developments 209
References 211

Index 215

Contributors

P.H.T. Beckett, University of Oxford, Agricultural Science Building, Oxford 0X1 3PF, Great Britain

J. Bouma, Wageningen Agricultural University, P.O.B. 37/6700 AA, Wageningen, The Netherlands

M.P.W. Farina, Specialist Scientist, Grain Crops Research Institute, South Africa

M.V. Fey, University of Natal, P.O. Box 375, Private Mail Bag, Pietermaritzburg 3200, South Africa

W.P. Miller, Department of Agronomy, University of Georgia, Athens, Georgia 30602, USA

M.A. Pavan, Instituto Agronomico do Parana, CX. Postal 1331, CEP 86.100, Londrina-PR, Brazil

I. Shainberg, Institute of Soils and Water, The Volcani Center, Bet Dagan 50-250, Israel

M.E. Sumner, Department of Agronomy, University of Georgia, Athens, Georgia 30602, USA

K.G. Tiller, Division of Soils, Commonwealth Scientific and Industrial Research Organization, Glen Osmond, South Australia 5064, Australia

Use of Gypsum on Soils: A Review*

I. Shainberg, M.E. Sumner, W.P. Miller, M.P.W. Farina, M.A. Pavan, and M.V. Fey

I. Introduction 2
A. Interest in Agricultural Uses of Gypsum 2
B. Availability of Gypsum 2
C. Aqueous Chemistry of Gypsum 4
II. Use of Gypsum on Dispersive Soils 6
A. Definition of Dispersive Soils 6
B. Effect of Solution Composition on Swelling and Dispersion of Clays 10
C. Gypsum and Hydraulic Conductivity (HC) Reclamation Of Sodic Soils 15
D. Gypsum, Rain Penetration, and Water Conservation 22
E. Gypsum and Soil Conservation 43
F. Gypsum and Water Penetration in Surface Irrigation 50
G. Gypsum and Seedling Emergence 52
H. Yield Response and Economics 53
III. Gypsum As an Ameliorant for Acid Subsoils 54
A. Introduction 54
B. Crop Response 59
C. Effects on Soil Properties 77
D. Studies in Pure Solutions 89
E. Column Leaching Studies 89
F. Explanation of the Amelioration by Gypsum 90
G. Practical Considerations 94
H. Adverse Effects of Phosphogypsum Use 95
I. Gypsum As a Nutrient Source 96
IV. Environmental Concerns in Phosphogypsum Use 97
V. General Conclusions 100
References 101

---

*Contribution from the Department of Agronomy, University of Georgia, Athens, Georgia 30602.

I. Introduction

A. Interest in Agricultural Uses of Gypsum

Gypsum is a relatively common mineral that is widely available in agricultural areas and has a number of specialized agronomic uses, principally as a Ca source on legumes and as a soil conditioner on sodic soils. Recent research has suggest that the utility of gypsum may extend to a greater range of soils and crops than previously acknowledged. The purpose of this review is to document both current practices and potential new uses of gypsum in agriculture, in order to establish the state of the art in current gypsum use, and to point out needed research in this important area.

With the need today to both enhance yields per unit of input in the harsh economic climate of industrialized agriculture and improve overall productivity in developing countries, gypsum use has been reevaluated in the light of recent findings on its effect on soil physical and chemical properties. These findings have significant implications for two important developing agricultural regions that have substantial soil-related limitations: the semiarid regions, exemplified by the Sahel region of Africa, and the equatorial tropics of Central and South America and Africa. In the former areas, poor rainfall distribution coupled with severe runoff and erosion problems have contributed to famine and food importation, while in the tropics soil infertility and acidity have restricted yields and associated agricultural development.

This review shows the potential of gypsum amendment to address both these diverse problems: the improvement of rainfall acceptance and lessening of runoff and erosion, and the amelioration of many problems associated with soil profile acidity limitations to plant growth. The effect of gypsum on soil crusting and infiltration, well known on sodic soils, has recently been extended to include a wide range of dispersive, low-Na soils such as occur in the southeastern United States. In this region, both improved infiltration and deeper rooting into acid subsoils result from the use of gypsum, and yields are thereby significantly enhanced.

Widespread, commercial-scale use of gypsum on dispersive and/or acid soils for field crops depends strongly on the availability and cost of gypsum, as balanced against yield responses obtained by such applications. Where gypsum is available at reasonable cost, significant advances in yields and overall efficiency may be possible on a variety of crops and soils.

B. Availability of Gypsum

Gypsum ($CaSO_4 \cdot 2H_2O$) and its dehydration products, the hemihydrate (plaster of Paris, $CaSO_4 \cdot 1/2H_2O$) and anhydrite ($CaSO_4$), are widely distributed minerals, found worldwide in sedimentary evaporite deposits (Hurlbut and Klein, 1971). In evaporating brines, gypsum is one of the first minerals to precipitate, and is typically found beneath rock salt deposits in marine salt domes (Murray, 1964). At high temperature (>42°C), anhydrite may precipitate before gypsum, but the dihydrate mineral is much more common, as anhydrite easily rehydrates to gypsum upon exposure to water at lower temperature. The hemihydrate may form in gypsiferous surface soil horizons at elevated temperatures (Doner and Lyn, 1977), but is rare geologically. Although relatively pure gypsum deposits are known, impurities of Ca and Mg carbonate and sulfate salts, as well as Fe oxides, are common.

Large gypsum deposits are found in the United States (notably in Arizona, New Mexico, New York, Texas, and Iowa), as well as in Canada, France, England, and the Soviet Union. The main commercial use of mined gypsum is for wallboard and as a cement additive, but significant amounts are marketed locally as a soil amendment (Hurlbut and Klein, 1971).

Other nongeologic sources of gypsum stem from industrial processes producing $CaSO_4$ by-products. The most important is the production of phosphoric acid from rock phosphate (apatite) by wet-process aciculation:

$$Ca_{10}(PO_4)_6F_2(s) + 10H_2SO_4 + 20H_2O \rightarrow 10CaSO_4{\cdot}2H_2O(s) + 6H_3PO_4 + 2HF$$

The phosphoric acid is used to manufacture high-analysis P fertilizers, while the by-product gypsum is collected as a waste product. This fine-grained, high-purity material, termed phosphogypsum, is produced in large quantities in Florida, Europe, and other locations with large phosphoric acid plants, and presents a serious disposal problem.

Other industries that produce waste sulfuric acids also commonly generate by-product gypsum by neutralization of this acid with hydrated lime:

$$Ca(OH)_2 + H_2SO_4 \rightarrow CaSO_4{\cdot}2H_2O$$

Many chemical processing and plating industries produce appreciable quantities of gypsum by this process that are often of high purity so long as near-stochiometric amounts of base are used to prevent contamination of the gypsum with other metals in the acid solution.

A potentially growing source of by-product gypsum is the capture of SO_2 from stack gases produced by fossil fuel-fired electric power generating plants. Environmental legislation is aimed at reducing this source of atmospheric acidity, and may mandate SO_2 "scrubbing" at some date. This may be accomplished by adding lime to the combustion chamber of boilers, or by passing the stack gases through a lime slurry, where the solvated SO_2 gas is oxidized to SO_4^{2-}:

$$2CaCO_3(s) + SO_2{\cdot}2H_2O(aq) \rightarrow CaSO_4{\cdot}2H_2O(s) + Ca^{2+}(aq) + 2CO_2(g) + 2H_2O$$

Efficiency as great as 95% has been achieved in pilot plants by this method, producing a high-purity product from which fly ash and other gas contaminants have been previously removed (Behrens and Hargrove, 1980). If SO_2 removal is implemented widely in this country and in Europe, hundreds of millions of tons of by-product gypsum would be produced annually.

Elemental analyses for a selection of mined and by-product gypsum samples are given in Table 1. The Nova Scotian gypsum commonly marketed in the

Table 1. Total analyses of selected gypsum samples

	Mined gypsum, Nova Scotia[a]	Phosphogypsum, Florida[b]
Major components (%)		
H_2O, chemical	15.9–20.2	14.4–18.8
CaO	31.2–34.7	26.9–31.1
SO_3	44.2–46.7	31.9–42.0
P_2O_5	ND	0.5–3.7
SiO_2 + insoluble	0.1–2.5	3.2–17.7
Al_2O_3	<0.01–0.13	<0.1–0.9
F	ND	0.2–0.8
Fe_2O_3	0.04–0.31	<0.05–0.2
MgO	0.03–0.70	<0.05
CO_2	<0.1–2.5	ND
Trace elements (mg kg^{-1})		
Arsenic	ND	2–8
Cadmium	ND	3–6
Copper	ND	7–11
Lead	ND	2–12
Selenium	ND	<0.5–2
Potassium	ND	20–86
Sodium	ND	150–4500

[a] From Cole (1930); ranges from 12 sample analyses.
[b] From May and Sweeney (1983); ranges from 5 sample analyses.
ND, Not determined.

United States is of high quality, although deposits are obviously quite variable in amounts of admixed impurities. The phosphogypsum samples contain various amounts of sand and clay residues carried over in phosphate rock processes, although generally this constitutes less than 5% by weight of the material. The by-product gypsum materials are typically of finer particle size than the mined samples, which are crushed to varying sizes before use. The effect of particle size on dissolution rate of soil-applied gypsum is an important factor in determining the effectiveness of gypsum amendment.

C. Aqueous Chemistry of Gypsum

Gypsum and its dehydration products are slightly soluble salts in aqueous solution, dissolving to an extent of approximately 2.5 g l^{-1}, or 15 m*M*. This level of solubility is a substantial contribution to the ionic strength of most soil solutions, yet is low enough to allow continued release of salt to the soil over a considerable time period. Other common Ca salts are either much less soluble ($CaCO_3$) or more soluble [$CaCl_2$, $Ca(NO_3)_2$]. Equilibrium solutions of several gypsum samples (Table 2) demonstrate the near-saturation levels reached during shaking with

Table 2. Dissolution of selected gypsum samples in water

Sample	Equilibrium Ca^{2+} concentration (m*M*)	Electric conductivity ($\mu S\ cm^{-1}$)	pH
Mined gypsums[a]			
U.S. Gypsum	15.2	1940	7.0
Gold Bond	16.7	1920	6.9
420	16.7	1950	6.7
By-product gypsums[b]			
Gulf Power	15.9	1890	6.8
Occidental	17.0	1940	4.7
Savannah	17.6	2360	6.7

[a] Names are trade products containing crushed mined gypsum.

[b] "Gulf Power" product is a desulfurization waste; "Occidental" is a phosphogypsum; "Savannah" is an industrial waste of unknown origin.

water. The alkaline pH levels of the mined samples reflect $CaCO_3$ impurities in the rock, while the slightly acid reaction of the phosphogypsum is a result of small amounts of free acid remaining in the by-product.

Gypsum solubility is significantly enhanced by complex formation. The $CaSO_4^0$ ion pair contributes to solubility in aqueous solution, as do reductions in activity coefficients from Debye–Huckel interactions. In Na salt solutions, the diverse ion effect increases solubility above that in pure water, because of both ionic strength effects and ion pairing (Nakayama, 1971). In saline soils, solubility is enhanced by the presence of soluble salts in the soil solution. Overall dissolution of gypsum in soils is promoted by exchange of Ca for other exchangeable ions, which may have a limited effect on raising equilibrium Ca levels by releasing diverse ions into soil solution.

In soil systems water is often in flux through the pores (during percolation) or over the soil surface (during runoff), and thus the rate of dissolution is an important parameter affecting the efficiency of gypsum amendment. Keren and Shainberg (1981) showed that by-product phosphogypsum was more rapidly soluble than mined gypsum, and was therefore more effective in improving infiltration rates of Israeli soils. Increasing particle size is a major factor limiting dissolution rates, largely through the decreased surface area of larger particles. The interaction of particle size and water flow velocity has been shown to determine resultant Ca^{2+} concentrations in gypsum-lined ditches (Kemper et al., 1975). Thus, to maintain relatively high concentrations of Ca in the thin film of water at the soil surface, the particle size of gypsum must be sufficiently small to promote rapid dissolution relative to the velocity of the runoff water. Similarly, the rate of movement of Ca in the soil profile via percolation is dependent on the same parameters.

The data in Figures 1 and 2 show dissolution curves for several gypsum samples. The Israeli samples (Figure 1) demonstrate the reduced rates of solubility of

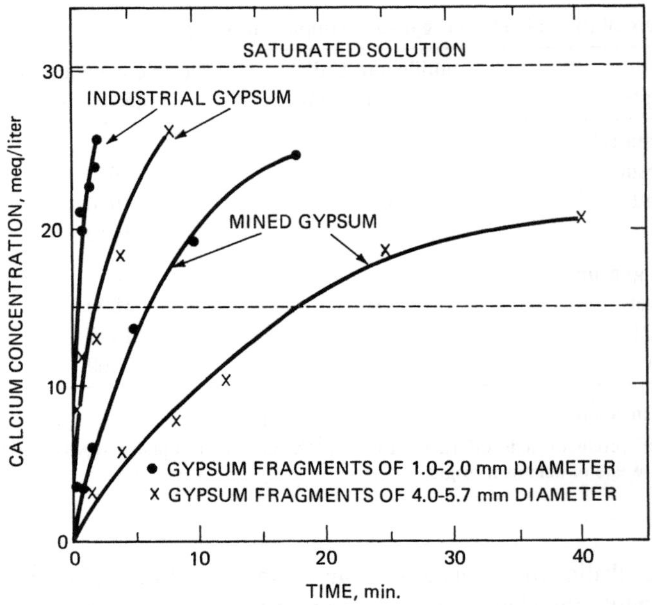

Figure 1. Calcium concentration versus time for industrial and mined gypsum samples for two fragment sizes (Keren and Shainberg, 1981).

mined gypsum compared to industrial (phosphogypsum), possibly because of $CaCO_3$ coatings on the geologic materials or their larger crystals. Particle size also shows an obvious effect on solubility, with significant differences between coarse sand and gravel-sized material. On very fine-grained gypsum ($<50\mu m$), differences in solubility kinetics were less pronounced among reagent-grade, by-product, and mined gypsum (Figure 2). Various sand-sized particle sizes, however, showed significantly reduced solubility over time. The importance of including silt-sized particles in surface-applied gypsum is obvious from these results: The immediate dissolution displayed by these fine particles will maintain high concentrations of electrolyte in percolating and runoff solutions.

II. Use of Gypsum on Dispersive Soils

A. Definition of Dispersive Soils

Many soils in semiarid to humid regions have unstable structure, which makes them susceptible to erosion and difficult to manage because of their tendency to disperse and develop compacted structure, particularly at or near the soil surface. Breakdown of soil aggregates at the soil surface leads to problems such as surface crusting, reduced water infiltration, high runoff and erosion, and restricted plant establishment and growth. Breakdown of aggregates occurs from slaking and clay

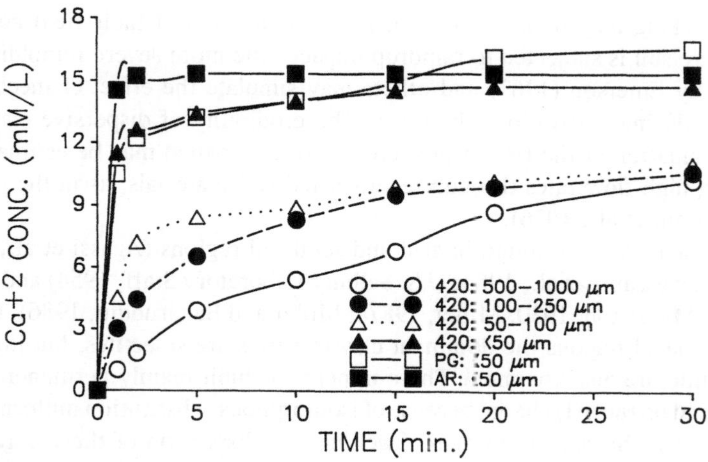

Figure 2. Dissolution kinetics of various kinds and particle sizes of gypsum in water (AR, reagent-grade $CaSO_4$; PG, phosphogypsum; 420, mined agricultural gypsum) (Miller, unpublished data).

dispersion when wetted. Aggregates susceptible to breakdown by these processes are also sensitive to mechanical stress, particularly when moist. Thus, raindrop impact and cultivation promote the development of compacted layers at the surface of soils with unstable structure.

Clay dispersion is defined as the formation of a stable suspension of particles in water. Dispersion is caused by the mutual repulsion between particles, which results from the presence of extensive negative electric fields surrounding them. Dispersion is enhanced by factors that increase the magnitude and extent of this field, such as the presence of highly hydrated monovalent cations (Na^+), increased pH, which activates variable charge by deprotonating hydroxyls on the clay surface, and low electrolyte concentration in solution. Flocculation is the opposite process, whereby the electric double layer is sufficiently compressed so that attractive (van der Waals) forces allow coagulation into microaggregates (Van Olphen, 1977). In field soils, flocculation is a necessary condition for the formation and stabilization of soil structure. The probable dispersive behavior of soils under different conditions may be estimated in various ways. A simple dispersion test developed by Emerson (1967) and modified by Loveday and Pyle (1973) is conducted by placing aggregates in a large volume of distilled water and measuring clay concentration in the suspension at a distance from the disintegrating aggregates. A distinction is made between spontaneous dispersion, where the aggregates in the suspension are not mixed, and mechanical dispersion, where the aggregates are shaken in water. Measurements of spontaneous dispersion reflect the behavior of the soil profile under an hydraulic head (hydraulic conductivity of the soil), or the behavior of a protected soil surface (mulch or permanent cover) during rainfall events. The extent to which soil particles disperse after gentle shaking may be used as a qualitative indication of their field behavior when bare soil is subjected to raindrop impact. The more severe remolding test as used by Emerson (1967) and others may simulate the effect of mechanical shearing during cultivation when wet. The erodibility of dispersive clay soils (e.g., evaluation of the risk of pipe erosion of clay dams) may be estimated by inducing high flow rates through recompacted soil materials, as in the pinhole test (Sherard et al., 1976).

Dispersive soils are found in arid and semiarid regions (Agassi et al., 1982, 1985a; Rengasamy et al., 1984; U.S. Salinity Laboratory Staff, 1954) and humid regions (Middleton, 1930; Miller, 1987a; Miller and Baharuddin, 1986). In soils from semiarid regions the dominant clay minerals are smectites, but illite and vermiculite are also common. These minerals exhibit mainly permanent (does not depend on the pH) charge because of isomorphous substitution uniformly distributed over the planar surfaces. Swelling and dispersion of these clays have been extensively studied (Bresler et al., 1982; McNeal and Coleman, 1966; McNeal et al., 1968; Quirk and Schofield, 1955; Shainberg and Letey, 1984; U.S. Salinity Laboratory Staff, 1954; Van Olphen, 1977). The predominant clay mineral in highly weathered soils from humid regions is kaolinite with sesquioxides. Most of the humid soils also contain organic matter. These materials exhibit variable (pH-dependent) charge as the result of reactive groups for which H^+ and OH^- are potential determining ions (Bowden et al., 1980). Whereas soil dispersion and swelling in smectitic soils are determined mainly by the presence of Na^+ on the exchange complex and only slightly by the pH of the system, soil dispersion in kaolinitic soils is predominantly determined by pH (Schofield and Samson, 1954) and the presence of dispersive anions derived from organic matter (Shanmuganathan and Oades, 1983a). The dispersivity of both types of soil is counteracted by electrolytes in the soil solution.

The favorable effect of exchangeable Ca^{2+} and the deleterious effect of exchangeable Na^+ on soil swelling and dispersion is well known. However, the critical level of exchangeable Na that separates dispersive (sodic) soils from nondispersive soils is not clear. The U.S. Salinity Laboratory Staff (1954), after studying many soil samples from the western United States, suggested an exchangeable sodium percentage (ESP) of 15 as the boundary between sodic and nonsodic soils, with soils of higher ESP being dispersive and suffering serious physical breakdown when wetted. Subsequently a modification by Bernstein (1974) took account of the influence of soil texture, with the critical ESP being 10 for fine- and 20 for coarse-textured soils. In Australia, serious physical limitations have been observed at ESP levels as low as 6 (Northcote and Skene, 1972). McIntyre (1979), studying the hydraulic conductivity (HC) of 71 soil samples from Australia, showed a continuous and initially very rapid decline in HC as ESP increased above zero, and proposed an ESP of 5 as the boundary between sodic and nonsodic soils.

This difference in critical ESP value is very important. Soils with ESP values greater than 15 are quite rare in semiarid and arid regions and are almost nonexistent in humid regions, while soils with low ESP values (1.0 < ESP < 5.0) are widespread in both semiarid and humid regions. Thus, the occurrence of dispersive and unstable soils that might respond to gypsum application is much wider than previously thought if one accepts the Australian definition. Moreover, a knowledge of the reasons for the difference in critical Na level between Riverside (United States) and Canberra (Australia) is necessary for an understanding of dispersive soils (Shainberg et al., 1981a; Shainberg and Letey, 1984). In the U.S. Salinity Laboratory, the HC of disturbed soil samples was measured using tap water whose electrolyte concentration was 5–10 mmol(+) l^{-1}. The lowest electrolyte concentration (EC) that McNeal and Coleman (1966) used in measuring the effect of solution composition on soil HC was 3.13 mmol(+) l^{-1}. At these electrolyte concentrations, relatively high ESP values were needed to cause a drop in HC (Shainberg et al., 1981a). Conversely, in Australia, the tap water used had an EC of only 0.7 mmol(+) l^{-1} (McIntyre, 1979), which is not sufficient to off set the dispersive effect of exchangeable Na at very low levels. Thus, when soils are exposed to rainwater (distilled water), susceptibility to low levels of exchangeable Na is enhanced, and is further exacerbated by raindrop impact, which causes mechanical in addition to chemical dispersion (Oster and Schroer, 1979; Agassi et al., 1985a).

In a recent publication, Rengassamy et al. (1984) proposed a scheme for predicting and classifying the dispersive behavior of surface layers of red-brown soils from southeastern Australia. In their study, surface soils with sodium adsorption ratio (SAR) > 3 would disperse spontaneously whereas those with SAR < 3 dispersed only after mechanical shaking. Because dispersion of the soil was dependent on both SAR and EC of the solution, both were included in their classification scheme. They divided the soils into three classes:

Class 1. Dispersive soils: soils that disperse spontaneously. These soils will have severe problems associated with crusting, reduced porosity, etc., even when subjected to minimal mechanical stress, for example, under zero tillage. Soils that spontaneously disperse under field conditions are nonsaline but sodic (SAR > 3.0).

Class 2. Potentially dispersive soils. Soils that disperse after mechanical shaking are potentially dispersive. Such soils are unstable if mechanically dispersed, for example, by intensive cultivation or raindrop impact. The EC required to keep these soils flocculated varies with their SAR and mineralogy. Soils from the A-horizon with SAR < 3 and very low EC (less than the flocculation value of the clay) and soils with SAR > 3 but with relatively high EC (exceeding the flocculation values of the clays) belong to this class (Class 2a).

Class 3. Flocculated soils. Soils that do not disperse even after mechanical shaking.

Whether and how soils respond to gypsum depends on their dispersion class. This response depends on the processes involved (HC, rain infiltration, erosion, etc.) and will be evaluated after a discussion of clay dispersion and swelling.

B. Effect of Solution Composition on Swelling and Dispersion of Clays

1. Na/Ca Distribution in Smectites

The diffuse double layer at clay surfaces consists of the lattice charge and compensating counterions, which reside in the liquid immediately adjacent to the particles. Counterions are subject to two opposing tendencies: (1) electrostatic attraction to negatively charged clay surfaces and (2) diffusion from high concentration at the surface of the particle to low concentration in the bulk of the solution. The result is an exponentially decreasing counterion concentration from the clay surface to the bulk solution. Divalent ions are attracted to the surface with a force twice as great as that of monovalent ions. Thus, in a divalent ion system, the diffuse double layer is more compressed. With an increase in the EC of the solution, the tendency of counterions to diffuse away from the surface is diminished and the diffuse double layer is further compressed. A complete description of diffuse double layer theory is presented elsewhere (Bresler et al., 1982; Van Olphen, 1977).

When two clay platelets approach each other, their diffuse counterion atmospheres overlap. Work must be done to overcome the electrical repulsion forces between the two positively charged ionic atmospheres. The electric double layer repulsion force, also called swelling pressure, can be calculated by means of the diffuse double layer theory, (Bresler et al., 1982; Van Olphen, 1977). The greater the compression of the ionic atmosphere toward the clay surface, the smaller the overlap of the atmospheres for a given distance between the particles. Consequently, repulsion forces between particles decrease with increasing EC and valence of the adsorbed ions. Because adsorbed Na ions form a diffuse layer, high swelling pressures develop between Na-montmorillonite platelets, and single platelets tend to persist in dilute solutions (Banin and Lahav, 1968; Shainberg et al., 1971). Conversely, low swelling pressures between Ca-clay platelets and electrical attraction forces between the exchangeable Ca ions and the negative clay surface (forces that are not considered in the diffuse double layer theory) prevent the complete swelling of Ca montmorillonite, even in distilled water, so that Ca platelets aggregate into tactoids (Blackmore and Miller, 1961) or quasicrystals (Aylmore and Quirk, 1959). Each tactoid consists of several (four to nine) clay platelets in parallel array, with an interplatelet distance of 0.9 nm. The exchangeable Ca ions adsorbed on the internal surfaces of the tactoids do not form a diffuse double layer. A diffuse ion layer is present only on the outside of these tactoids (Blackmore and Miler, 1961), and the clay behaves as a system with a much smaller surface area.

In earlier studies of mixtures of monovalent and divalent ions in smectite systems, the two ions were pictured as being mixed at random throughout the exchange complex (Bresler, 1970), which means that for any ratio of Na to Ca there are as many Na ions inside the tactoids as on their external surfaces. Mering and Glaeser (1954) hypothesized that "demixing" of the cations occurs so that a few interlayer spaces contain mainly Na ions and others mainly Ca ions. McAtee

Figure 3. Dependence of electrophoretic mobility (curve A) and the relative size (curve B) of montmorillonite particles on the exchangeable sodium percentage (ESP). The relative size is expressed in units of the slope of Einstein's equation for the viscosity of the suspension (Bar-On et al., 1970).

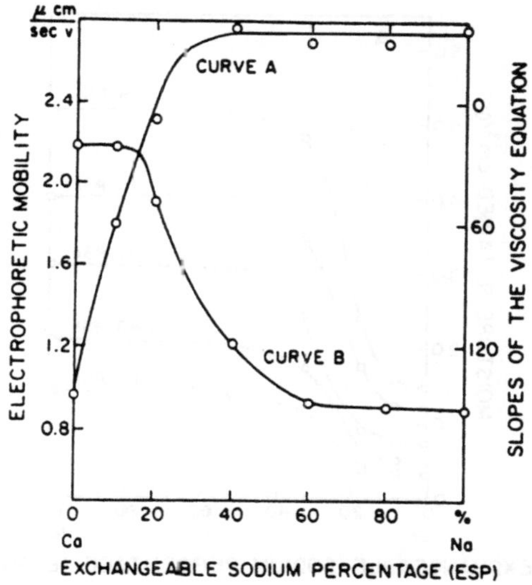

(1961) also inferred, from x-ray diffraction patterns of clay saturated with mixtures of mono- and divalent ions, that "demixing" had occurred.

Using viscosity and light transmission measurements, Shainberg and Otoh (1968) studied the size and shape of montmorillonite particles saturated with a mixture of Na and Ca ions in the adsorbed phase. They found (Figure 3, curve B) that the introduction of a small percentage of Na into the exchange complex of Ca tactoids was not sufficient to break the tactoid apart, but more Na (>20%) resulted in tactoid breakdown. The platelets were completely separated when 50–60% of the adsorbed Ca had been replaced by Na.

The location of the adsorbed ions in montmorillonite saturated with a mixture of mono- and divalent cations was estimated by Bar-On et al. (1970) through the measurement of the electrophoretic mobility of clay particles in suspension (Figure 3, curve A). The addition of a small amount of exchangeable Na- to Ca-saturated clay had a considerable effect on the electrophoretic mobility. The fact that the size of the Ca tactoids was not affected by a low ESP (Figure 3, curve B), whereas the electrophoretic mobility increased rapidly, supports the "demixing" model. At low ESP values, the adsorbed Na concentrates on the external surfaces of the tactoids and their electrophoretic mobility increases rapidly. As the ESP of the system increases, Na penetrates into the tactoids and causes their disintegration.

2. Swelling and Dispersion of Na/Ca Smectites and Illites

The volume of water retained by montmorillonite as a function of ESP for various consolidation pressures is presented in Figure 4. Clay swelling is affected only slightly by an increase in ESP at values less than 15. These results compare

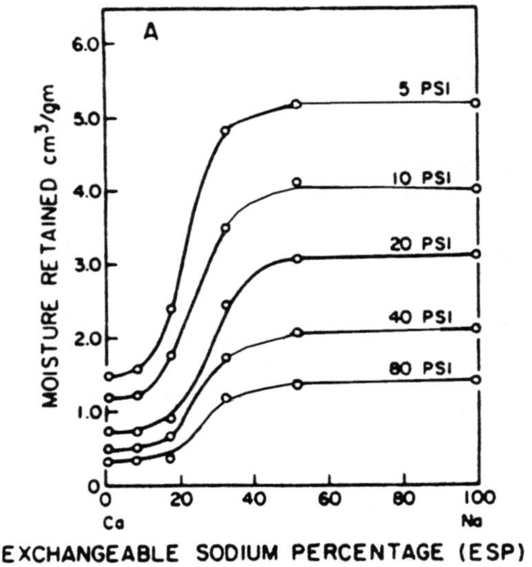

Figure 4. Water retention as a function of ESP and pressure applied on montmorillonite. (Reprinted with permission from I. Shainberg, I. Bresler, and Y. Klausner, Studies on Na/Ca montmorillonite systems. I. The swelling pressure, *Soil Sci.*, 111, pp. 214–219, © by Williams & Wilkins, 1971.)

favorably with those of McNeal and Coleman (1966) on soil clays, in which replacing about 15% of adsorbed Ca with Na had little effect on clay swelling. With a further increase in ESP, a very sharp increase in macroscopic swelling was found for two montmorillonitic soils.

In a stable clay suspension, dispersed particles frequently collide because of Brownian motion, but separate again because of diffuse double layer forces. When salt is added to the clay suspension, particles adhere upon collision, forming flocs that settle. The suspension is then separated into bottom sediment and particle-free supernatant liquid. The minimum EC that causes flocculation is referred to as the flocculation value, which depends on counterion valency. Flocculation values for Na- and Ca montmorillonite are 12 mmol(+) l^{-1} NaCl and 0.25 mmol(+) l^{-1} $CaCl_2$, respectively (Van Olphen, 1977), and for Na- and Ca illites they are 40–50 mmol(+) l^{-1} NaCl and 0.25 mmol(+) l^{-1} $CaCl_2$, respectively (Arora and Coleman, 1979).

Oster et al. (1980) found that a small increase in ESP of a Na/Ca montmorillonite had a considerable effect on the flocculation value. Similar effects of exchangeable Na on the electrophoretic mobility of Ca montmorillonite were reported (Figure 3, curve A). According to Verwey and Overbeek's stability theory (Van Olphen, 1977), these two properties are interrelated and both are explained by the "demixing" model, which postulates that Na is the predominant cation on the external surfaces.

The effect of exchangeable Na on the flocculation value of Ca-illite suspensions follows more closely a simple linear relationship (Oster et al., 1980), which suggests that "demixing" in Na/Ca illite is not as pronounced as in Na/Ca montmorillonite. However, illite at low ESP was more dispersed than the corresponding montmorillonite (Oster et al., 1980). Illite particles consist of a number of platelets stacked to a thickness of 10 nm, with a specific surface area of 120 $m^2 g^{-1}$ (for Fithian illite) as compared with a specific surface area of 750 $m^2 g^{-1}$ for montmorillonite. Electron micrographs reveal that illite particles have irregular surfaces and that the planar surfaces are terraced. Upon close approach, the unavoidable mismatch of the terraces leads to poor contact between the edges and the surfaces, leading to smaller edge-to-face attraction forces and, consequently, a higher flocculation value for Na illite compared with Na montmorillonite.

3. Dispersion and Flocculation of Kaolinites

Schofield and Samson (1954) studied the behavior of kaolinite in suspension and found that a pure Na kaolinite was flocculated at pH < 7 under conditions that dispersed illite and montmorillonite. The acidified kaolinite retained exchangeable Na, which showed that the crystals were negatively charged. When shaken in dilute NaCl solutions, Cl was positively adsorbed, which proved that the edges were positively charged. The attraction of the positive charges on the edges for the negative charges on the planar surfaces was regarded as the cause of the flocculation, which occurred in the absence of salt. Deflocculation of the salt-free suspension occurred on the addition of NaOH. At pH > 8, the crystals negatively adsorbed Cl, which showed that the edges were no longer positively charged. Thus, eliminating the positive charge at the edges dispersed the clay.

Deflocculation of kaolinite suspension could also be brought about by adding small amounts of sodium oxalate, sodium pyrophosphate, sodium polymetaphosphate, sodium alginate, and sodium bentonite (Schofield and Samson, 1954; Durgin and Chaney, 1984). The negatively charged polyanions are adsorbed on the positively charged edges, and thus prevent edge-to-face flocculation. A rapid and large increase in flocculation value of Na kaolinite was observed with increasing amounts (up to 30%) of Na smectite (Arora and Coleman, 1979; Schofield and Samson, 1954). They attributed this strong effect to the disposition of smectite platelets on positively charged edges of kaolinite. Similarly, Frenkel et al. (1978) found that the HC of kaolinitic soil from North Carolina (with 10.6% clay) was not affected by 20% Na on the exchange complex. However, when the soil was mixed with 2% montmorillonite (about 20% of the clay content), the mix was very susceptible to dispersion. The low CEC of dispersive Georgia soils indicated to Chiang et al. (1987) that the montmorillonite content in these soils could not be much more than 10% and could not account for the dispersion they observed. Thus, they concluded that impurities other than montmorillonite were masking the edge charge of kaolinite particles.

The effect of natural organic molecules on kaolinite dispersion was studied by Durgin and Chaney (1984) and Shanmuganathan and Oades (1983a). Durgin and Chaney (1984) studied the organic constituents of water extracts from Douglas fir roots that cause kaolinite dispersion. They found that aliphatic and aromatic carboxylic acids and organics with molecular weights greater than 10^4 were most effective in causing dispersion. Similarly, Shanmuganathan and Oades (1983a,b) distinguished three groups of anions in order of effectiveness with respect to clay dispersion: phosphate = fluvate > citrate = oxalate = tartarate > salicylate = lactate, which was the same order as for the amount of anion adsorption by the soil. Specific adsorption of the anions on the clay edges makes the clay more negative and thereby promotes dispersion.

4. Mechanism of Flocculation

Association between the positive edge surface of one particle and the flat negative oxygen surface of another was the mechanism proposed by Schofield and Samson (1954) to explain and describe the flocculation of kaolinite.

Van Olphen (1977) postulated similar positively charged sites on montmorillonite edges and edge-to-surface attraction to explain montmorillonite flocculation. The edge-to-face attraction hypothesis is supported by the following observations: (a) The flocculated Na-montmorillonite gel contains about 250 g of water per gram of clay, which corresponds to a film thickness of 330 nm. This thickness exceeds the range of diffuse double layer forces and/or van der Waals attraction forces. Thus only edge-to-face attraction forces are possible. (b) Addition of small amounts of Na polymetaphosphate [$(NaPO_3)_{13}$–Calgon] to Na-montmorillonite suspensions increased very sharply the flocculation value of the suspension. Chemisorption of the polyanion at the edges, and reversal of the positive edges charge to negative, may explain the effect of the dispersant on flocculation. (c) The flocculation value of Na montmorillonite increases with increasing pH (Goldberg and Glaubig, 1987). With increase in pH, the charge on the edges becomes negative, and the suspension becomes more stable.

However, several experimental observations oppose and contradict the edge-to-face mechanism in smectites. If edge-to-face association is prevented, the open structure of the gel should collapse. Thus the gel volume of flocculated Na montmorillonite containing polymetaphosphate or at high pH should be small. Experimentally, the open structure of the clay gel was maintained (Keren et al., 1988; Oster et al., 1980). Thus, another association mode was suggested by Keren et al. (1988), based on the fact that the measured charge density at a smectite surface is only an average value; the actual charge density is quite heterogeneous (Stul and Mortier, 1974) with regions on the clay where the charge density is twice the average value and is similar to the charge density on vermiculite. On surfaces with high charge density (vermiculite), there is a net electrostatic attraction force between the platelets (Norrish, 1954). Because the montmorillonite platelets are long, flexible, and have the capability of bending (Shomer and Mingelgrin, 1978), one platelet may simultaneously form a cohesive junction point with other platelets at several locations of high charge densities to form an open network with high gel volume. Thus, at neutral and low pH, edge-to-face association is the dominant mode of association in Na-montmorillonite gel. At high pH or in the presence of polyanions that neutralize the charge on the edge, face-to-face association in regions with high charge densities is taking place (Keren et al., 1988). To enable the electrostatic attraction between regions of high charge densities, the diffuse double layer must be further compressed, and higher EC is needed to coagulate the clay.

C. Gypsum and Hydraulic Conductivity (HC): Reclamation of Sodic Soils

1. Introduction

Soil permeability to water depends on both the ESP of the soil and the EC of the percolating solution, tending to decrease with increasing ESP and decreasing EC (McNeal and Coleman, 1966; McNeal et al., 1968; Quirk and Schofield, 1955). Soil HC can be maintained even at high ESP levels provided that the EC of the percolating water is above a critical (threshold) level (Quirk and Schofield, 1955). Conversely, when very good quality water is used, even an ESP value of 5 caused a two-order-of-magnitude decrease in the HC of soils (McIntyre, 1979).

Even in arid and semiarid regions where irrigation is practiced, a rainy season occurs (Shalhevet and Kamburov, 1976). Although the EC in most irrigation waters is greater than 5 mmol(+) l^{-1} (0.5 dS m^{-1}) (Shalhevet and Kamburov, 1976), rain EC is less than 1 mmol(+) l^{-1} (<0.1 dS m^{-1}). Thus, the deleterious effect of exchangeable Na is more evident during the rainy season than during irrigation. Moreover, some irreversible deterioration in soil physical properties that cannot be repaired during the irrigation season may occur at low ESP values during the rainy season.

Gypsum is the most commonly used amendment for sodic soil reclamation, primarily because of its low cost. Gypsum added to a sodic soil can increase permeability by means of both EC and cation-exchange effects (Loveday, 1976). The relative importance of the two effects is of interest for several reasons. If the electrolyte effect is sufficiently great to prevent clay dispersion and swelling, surface application of gypsum may be worthwhile. In this case, the amount of gypsum required depends on the amount of high-quality water applied and the rate of gypsum dissolution. It is somewhat independent of the amount of exchangeable Na in the soil profile. Conversely, in soils where the EC effect is of lesser importance than cation exchange, the amount of gypsum required depends on the amount of exchangeable Na in a selected depth of soil. The cation-exchange process has formed the basis of several "gypsum requirement" tests (Dutt and Terkeltoub, 1972; Oster and Frenkel, 1980; U.S. Salinity Laboratory Staff, 1954), whereas EC effects have generally been neglected.

In this section, the effect of electrolytes on soil HC, differences between soils in their response to sodic conditions, and the possibilities for recognizing gypsum responsive soils are discussed. A distinction is made between HC and infiltration rate (IR) of soils. Whereas HC measurements are made under conditions in which the soil surface is not disturbed, surface disturbance (beating action of raindrops, stirring effect of flowing water) is possible under IR measurements. Disturbance of the soil surface during water infiltration renders the surface more sensitive to low sodicity (ESP < 5), as mechanically disturbed soil disperses at lower sodicity than undisturbed soil (Emerson, 1967; Rengasamy et al., 1984). In addition, the EC of the soil solution at the soil surface is determined predominantly by the applied water, whereas that in the soil profile depends also on the rate of dissolution of the soil constituents. Thus in a calcareous soil the EC in the soil solution in the profile will seldom be less than 2 mmol l^{-1} even when

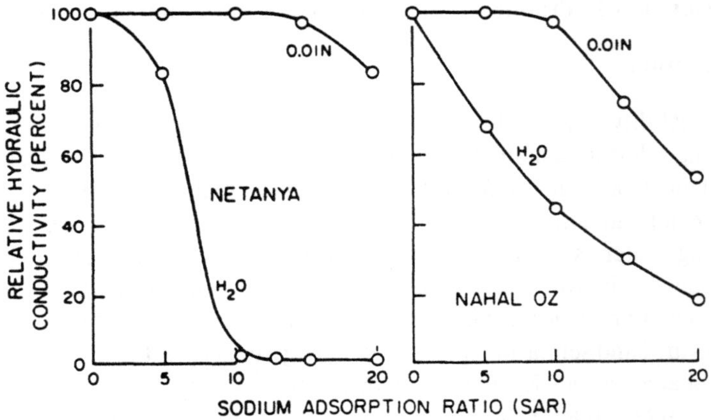

Figure 5. Hydraulic conductivities of a sandy loam (Netanya) and a silt loam (Nahal Oz) soil as a function of SAR and concentration of leaching solutions (Felhendler et al., 1974).

leached with distilled water. As we shall see, this electrolyte concentration is enough to prevent the adverse effect of low ESP, and the HC of the soil profile is susceptible only to ESP values exceeding 5.

2. Effect of Electrolyte Concentration on the HC of Smectitic Soils

The permeability of a soil to water depends on the ESP of the soil and on the EC of the percolating solution, and thus Quirk and Schofield (1955) developed the concept of "threshold concentration." This is the EC of the percolating solutions that causes a 10–15% decrease in soil permeability at a given ESP value.

At EC values of the percolating solution greater than 3 mmol(+) l^{-1}, soils responded differently to the same combination of EC and ESP; thus, a unique threshold concentration exists for each soil (McNeal and Coleman, 1966). When the salinity of the soil solution exceeds 3 mmol(+) l^{-1}, the susceptibility of soil to sodic conditions increased with an increase in (a) clay content (McNeal et al., 1968; Frenkel et al., 1978), (b) swelling clay, especially montmorillonite (McNeal and Coleman, 1966), and (c) soil bulk density (Frenkel et al., 1978). Soils rich in sesquioxides are less susceptible to sodic conditions (McNeal et al., 1968).

Felhendler et al. (1974) measured the HC of two montmorillonite soils (a sandy loam and a silt loam) as a function of the SAR and EC of the percolating solution and found that both soils were only slightly affected by the SAR (up to 20) of the percolating solution as long as the EC exceeded 10 mmol l^{-1} (Figure 5). However, when the percolating salt solution was displaced by distilled water, simulating rainfall, the response of the two soils differed drastically. The HC of the calcareous silt loam dropped to 42 and 18% of the initial value for soils with ESP values of 10 and 20, respectively. The HC of the noncalcareous sandy loam soil dropped to 5 and 0% of the initial value for the same respective con-

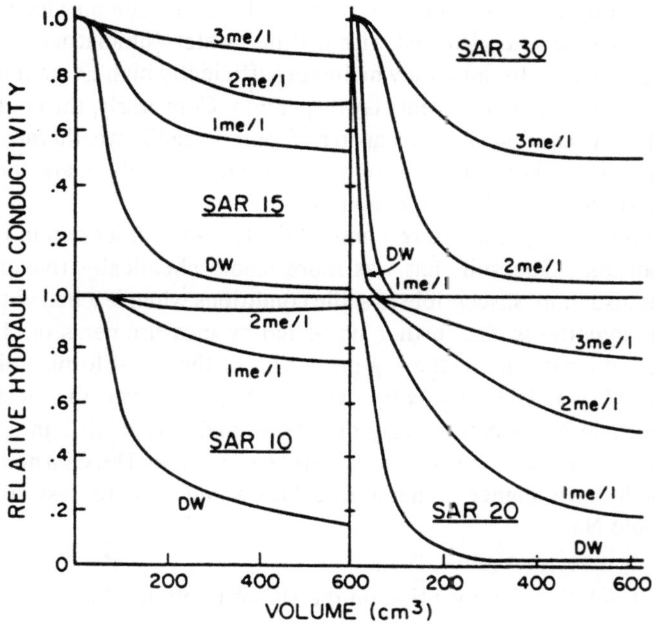

Figure 6. Relative hydraulic conductivities of a Fallbrook soil–sand mixture, equilibrated with 0.01 *N* solutions of SAR 10, 15, 20, or 30, and leached with distilled water (DW) or salt solutions of 1, 2, or 3 meq l^{-1} (Shainberg et al., 1981a).

ditions. Felhendler et al. (1974) also noted that the clay in the sandy loam soil was mobile and appeared in the leachate, whereas no clay dispersion occurred in the silt loam soil. An explanation of why the sandy loam was more susceptible to sodic condition than the silt loam soil was given by Shainberg et al. (1981a), who studied the effect of displacing 0.01 *N* solutions on SAR 10, 15, 20, and 30 with distilled water or solutions of 1, 2, and 3 mmol(+) l^{-1} (Figure 6). Even a low ESP was enough to appreciably reduce the HC of the Fallbrook soil when it was leached with distilled water. An EC of 2 mmol(+) l^{-1} in the percolating solution prevented the adverse effect of ESP 10 on the HC of this soil. The adverse effect of 15% Na on the exchange complex was prevented by a solution of 3 mmol(+) l^{-1}.

Young soils may release 3–5 mmol(+) l^{-1} of Ca and Mg into the solution as a result of the dissolution of plagioclase, feldspars, hornblende, and other minerals (Rhoades et al., 1968). The solution composition of a calcareous soil at a given ESP, when placed in contact with distilled water, is also in the range of 2–3 mmol(+) l^{-1} (Shainberg et al., 1981b). The sensitivity of sodic soils to solutions of low EC led Shainberg et al. (1981a,b) to hypothesize that a major factor causing differences among various sodic soils in their susceptibilities to decreases in HC when leached with low electrolyte water was their rate of salt release by mineral dissolution. Mineral dissolution determines the EC of percolating solutions. Sodic soils containing minerals that readily release soluble electrolytes will not easily disperse when leached with distilled water (simulating rainwater) at moderate ESP values because they maintain sufficiently high EC in the soil solution (>3 mmol(+) l^{-1}) to prevent clay dispersion. Conversely, the concentration of salt in the soil solution of soils that do not contain readily weatherable minerals will be below a threshold concentration (the flocculation value), and the soils will be more susceptible to clay dispersion.

This hypothesis explains the response of the two soils described in Figure 5 to sodic conditions. The soil that was more stable chemically (Netanya sandy loam), was also more susceptible to sodic conditions when leached with distilled water. The hypothesis was further supported by measurements of clay dispersivity, HC, and mineral weathering properties of three California soils (Shainberg et al., 1981b). The chemically stable Fallbrook soil was the most sensitive to exchangeable Na effects on clay dispersion and loss in HC, in spite of the presence of sesquioxides and kaolinite in its clay fraction. The Gila and Pachappa soils, which have higher mineral dissolution rates, were less affected by exchangeable Na.

3. Effect of Solution Composition on the HC of Kaolinitic Soils

Differences of opinion can be found in the literature on the effect of kaolinite clay on the HC of soils leached with sodic water. McNeal and Coleman (1966) and Yaron and Thomas (1968) concluded that the most and least labile soils were those high in 2:1 layer silicates (especially montmorillonite) and those high in kaolinite and sesquioxides, respectively. El-Swaify and Swindale (1969) found a negligible effect of exchangeable Na, even in the absence of salinity, on the HC of tropical soils whose clay fractions were dominated by kaolins, iron oxides, amorphous silicates, and gibbsite. McNeal et al. (1968) found that the "stability" of HC in Hawaiian soils under high-Na and low-salt conditions was greatly reduced by partial removal of free iron oxides, pointing to the cementing action of iron oxides in preventing dispersion. Velasco-Molina et al. (1971), concluded that, in the virtual absence of electrolyte, the order of soil dispersion at a given ESP was montmorillonitic > halloysitic > mica. Elgabaly and Elghamry (1970) found that the HC of ground and sieved kaolinitic systems with ESP > 10 decreased rapidly when leached with distilled water. Frenkel et al. (1978) concluded that although kaolinitic was less sensitive than montmorillonitic soil at low EC, its HC was reduced markedly, even at an ESP of 10, when leached with distilled water. Frenkel et al. (1978) and Chiang et al. (1987) concluded that soil behavior depended on the type of mineral that is present in combination with kaolinite. Soil kaolinites containing small amounts of montmorillonite and mica were dispersive (Arora and Coleman, 1979; Frenkel et al., 1978; Schofield and Samson, 1954). This phenomenon has been ascribed to the breakup of the edge-to-face particle association of kaolin structure by the adsorption of negatively charged montmorillonite faces on the positively charged kaolin edges.

In kaolinitic soils, pH also affects dispersion because of the variable nature of the positive charge. Peele (1936) showed that permeability and aggregate size of a Cecil series soil decreased as lime was added. Arora and Coleman (1979) found that raising the pH of a Georgia kaolinite from 7 to 8.3 created more dispersion than in any of the other samples including smectites, illites, and vermiculites. Studying the effect of EC, ESP, and soil pH on the HC of three kaolinitic soils from the southeastern United States, Chiang et al. (1987) showed that the HC of the Cecil soil was very sensitive to changes in EC, Na content, and pH, while the Davidson and Iredell soils were well flocculated and unaffected by changes in EC at ESP < 3. Thus, kaolinitic soils, being dispersive, will also respond to gypsum amendments through increased EC in the soil solution.

4. Reclamation of Sodic Soils with Gypsum

Gypsum is by far the most commonly used amendment for sodic soil reclamation because of its low cost, availability, and ease of handling. Its value has been known for a long time; Hilgard (1906), for example, discussed its role in preventing deflocculation, thus allowing leaching to proceed, and Kelley and Arany (1928) reported successful reclamation with gypsum. A great many published results have followed, so that we now have a good general understanding of its role in the reclamation process.

The basis of the process has already been outlined. Gypsum dissolves to provide a level of electrolyte in the soil solution that maintains sufficient permeability to allow water entry into and through the soil profile and at the same time provides Ca for exchange with Na. Potentially achievable EC, with the soil solution saturated by gypsum, ranges from 15 to 133 mol m^{-3} (30–266 mmol(+) l^{-1} as the ESP varies from 0 to 40 (Oster, 1982). Actual concentrations achieved may range from near saturation, as was apparently the case in the field experiment of Dutt and Terkeltaub (1972), to much less, of the order of one third to one half saturated, that is, 12–15 mmol(+) l^{-1} at the lower end (Quirk and Schofield, 1955; Chaudhry and Warkentin, 1968), and are influenced by experimental conditions. Gypsum directly prevents swelling and dispersion, and indirectly increases porosity, structural stability, hydraulic properties, soil tilth, drainage, and leaching, and reduces dry soil strength.

With regard to soil tilth, gypsum-induced improvements include a wider range of water contents at which tillage can be performed, and greater ease and effectiveness of tillage for seedbed preparation and for weed control (Milthorpe and Newman, 1979). Reduced tillage and land grading costs should result, as well as greater flexibility in date of sowing and harvesting and more uniform water entry at irrigation.

Surface crust strength is very largely dependent on its moisture content and, because gypsum treatment slows the rate of surface drying (Loveday and Scotter, 1966), the rate of crust development and final strength will be affected. These surface strength effects have often been recorded in terms of marked improvements in seedling emergence and establishment and in reductions in modulus of rupture and resistance to penetration (Loveday and Scotter, 1966).

Water penetration and storage benefits from gypsum have been described for a variety of soils throughout the world. In a ponding experiment on a brown sodic clay, McIntyre et al. (1982) found that, without gypsum, 292 mm infiltrated in 379 days of ponding, wetting the profile to 2.1 m; with gypsum at 10 t ha^{-1}, 605 mm infiltrated in 145 days, enough of which passed deeper than 2 m to raise the groundwater level. Data from this experiment clearly demonstrated the presence of a region of low HC in the upper profile that was ameliorated by the addition of gypsum.

The increased leaching of soluble salts, resulting from the extra profile drainage after gypsum application, has been documented on many occasions. In the experiment just mentioned, McIntyre et al. (1982) found depths of leaching of Cl, beyond which accumulation occurred, to be 1.0 m without gypsum and 2.8 m with gypsum. The expected outcome of these soil effects is improved crop and pasture yields.

a. Recognizing Gypsum-Responsive Soils: Significance of the Electrolyte Effect

In order to take advantage of the fact that worthwhile responses by both soil and crop may be obtained from quite modest gypsum application rates via the electrolyte effect, the first need is to be able to recognize a gypsum-responsive soil. The second need is to determine the "gypsum requirement," namely, the application rate that should effect "permanent" reclamation of the root zone profile. As pointed out earlier, this rate is often beyond the means of farmers in any one year, so that reclamation in stages is generally recommended (Loveday, 1984).

Shainberg et al. (1982) studied the relative significance of the electrolyte and reclamation effects on the HC of sodic soils (Figure 7). The relative importance of the electrolyte effect was estimated by comparing the HC of the sodic soils when reclaimed by equivalent amounts of $CaCl_2$ and $CaSO_4$. Leaching in all cases was with distilled water, simulating rainwater. Although the exchange reclamation was similar with both amendments, the highly soluble $CaCl_2$ replaced the exchangeable Na and was leached out of the soil profile rather quickly. The slightly soluble gypsum, on the other hand, was slowly released into the displacing water. Depending on the rate of water application and gypsum dissolution, sufficient EC could be obtained to prevent dispersion and accompanying reductions in soil HC.

The relative importance of the electrolyte and exchange effects introduced by gypsum depends on the properties of the soils (Shainberg et al., 1982). The most relevant property is the ability of the soil to release electrolytes when leached with distilled water. In Figure 7, the Golan soil, a highly weathered Rhodoxeralf, occurs under 1,000 mm of annual rainfall and is $CaCO_3$ free, while the Nahal-Oz soil, a calcareous Haploxeralf, is from a region with 350 mm of rainfall. The noncalcareous Golan soil was very sensitive to the type of Ca amendment added. Complete sealing of the soil eventually took place in the $CaCl_2$ treatments, but high HC was maintained in the gypsum treatments. Conversely, for the calcare-

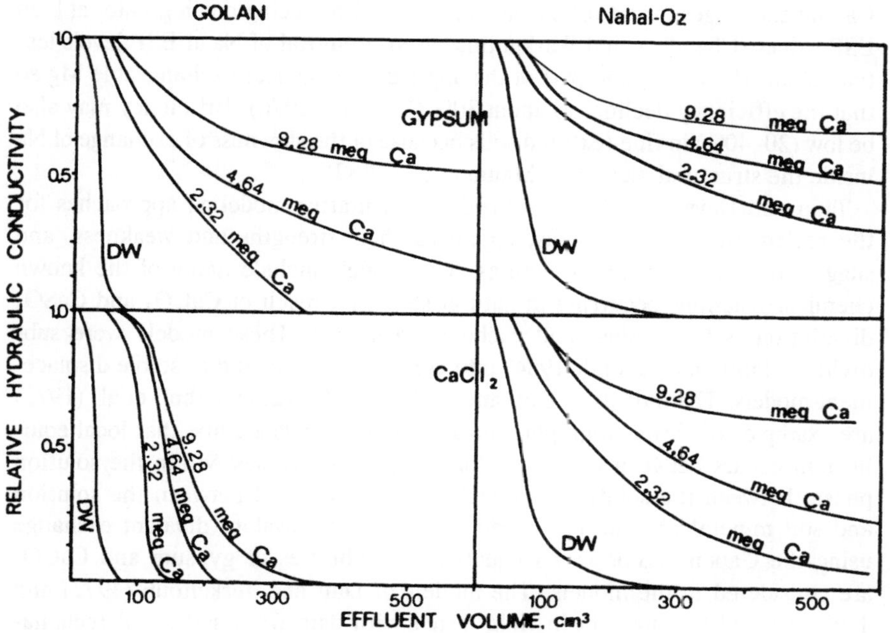

Figure 7. Relative hydraulic conductivities of the Golan and Nahal-Oz soils as a function of cumulative effluent volume, showing the effects of varying $CaSO_4$ and $CaCl_2$ applications (Shainberg et al., 1982).

ous Haploxeralf, no difference between the two amendments was observed. Because Na replacement was similar for both $CaSO_4$ and $CaCl_2$ (when applied in equivalent amounts), the differences in HC response of the two soils to the type of amendment were attributed to an electrolyte effect. The Golan soil, which does not have the potential to release appreciable amounts of electrolyte, is sensitive to low ESP values. Thus the release of electrolytes by gypsum was essential in order to maintain high HC. Because the $CaCO_3$ in the Nahal-Oz soil dissolved readily during leaching with distilled water, the electrolyte effect from gypsum was not additionally beneficial (Figure 7).

The beneficial effect of gypsum in preventing clay dispersion may particularly manifest itself at the soil surface. Calcareous soils with moderate ESP levels maintain enough EC in the soil profile to prevent HC decrease. However, the EC at the soil surface may be insufficient to prevent physical deterioration during raindrop impact.

b. Modeling and Calculating Gypsum Requirement

The amount of gypsum needed to reclaim a sodic soil, the "gypsum requirement" (U.S. Salinity Laboratory Staff, 1954), is inter alia a function of the depth of soil to be reclaimed, the initial content of exchangeable Na, and final ESP being sought. The efficiency and rate of exchange, namely, the percentage of applied

Ca that exchanges for adsorbed Na, varies with ESP, being much greater at high ESP values (Chaudhry and Warkentin, 1968). Removal of Na at ESP levels less than about 10 is slow, and part of the applied Ca displaces exchangeable Mg so that the efficiency declines to about 30% (Loveday, 1976). Efficiency may also be low (20–40%) in fine-textured soils because of the slowness of exchange of Na inside the structural elements (Manin et al., 1982).

Tanji and Deveral (1984) described representative modeling approaches for the reclamation of sodic soils, appraised their strengths and weakness, and suggested future research directions. The models include many of the known chemical reactions occurring in natural soils as a result of $CaCO_3$ and $CaSO_4$ dissolution, salt leaching, and exchange reactions. These models were subdivided (Tanji and Deveral, 1984) into chromatographic and miscible displacement models. The models of Dutt and Terkeltoub (1972) and Tanji et al. (1972) are examples of chromatographic models. Both models assume that local equilibrium occurs between the plates for chemical processes within the solution phase, between the solution and exchanger phases, and between the solution and soil mineral phases. The models compute monovalent–divalent exchange using the Gapon and/or Davis equations. Solubilities of gypsum and $CaCO_3$ are considered in the models. The models of Dutt and Terkeltoub (1972) and Tanji et al. (1972) have been field tested with data from sodic soil reclamation leaching plots. The former model was validated with observed ESP data from the 15- to 30-cm soil depth from 11 plots. The latter model was validated with observed data from nine soil depths between 0 and 270 cm for total soluble cations and SAR from 3 plots. The differences between measured and computed results were found to be no more than the horizontal variations typically found in salt-affected lands.

Both models provide powerful tools for quantitative predictions of water and gypsum required to reclaim soil profiles to prescribed levels of salinity and ESP. However, assumptions must be made about the desired level of reclamation: To what level must ESP be reduced, and to what depth of soil is the ESP reduction required? Our discussion in previous sections suggests that there is no unique level of ESP that can be regarded as critical. The soil surface, being very susceptible to sodicity, may tolerate only very low ESP (<3), whereas at depths below the cultivated layer, ESP > 15 may be tolerable. Thus our objective in reclaiming sodic soils might be only to spread gypsum at the soil surface. Dissolution of gypsum at the soil surface will release enough electrolyte to prevent clay dispersion and HC decline both at the surface and within the soil profile. Thus, frequent application (once a year, before the rainy season) of small amounts of gypsum, preferably phosphogypsum, which dissolves readily, may be the optimal practice in gypsum reclamation.

D. Gypsum, Rain Penetration, and Water Conservation

1. Introduction

The IR is defined as the volume flux of water flowing into the profile per unit surface area of soil, and has the dimension of velocity. In general, soil infiltration capacity is initially high, particularly when the soil is initially dry, but it tends to decrease monotonically until it asymptotically approaches a constant rate—the final or steady-state IR. If water delivery rate to the surface is smaller than the soil infiltration capacity, water infiltrates as fast as it is delivered and the supply rate determines the IR. When delivery rate exceeds soil infiltration capacity, the latter determines the actual IR, and the process becomes surface controlled.

In soils having stable surface structures, decreases in infiltration capacity result from the inevitable decrease in the matric suction gradient that occurs as infiltration proceeds (Baver et al., 1972; Hillel, 1980). In the initial wetting process, the surface is saturated but the soil below is dry and the matric suction gradient is very steep. As the wetted zone deepens, this gradient is gradually reduced.

Decreases in soil infiltration capacity from an initially high rate can also result from gradual deterioration of soil structure and the formation of a surface crust. When a crust of very low HC is formed at the soil surface, its reduced permeability determines the IR of the soil (Baver et al., 1972; Hillel, 1980; Morin and Benyamini, 1977).

Numerous formulations have been proposed in repeated attempts to express the IR as a function of time or of the total quantity of water infiltrated into the soil (Baver et al., 1972; Hillel, 1980). The equation most pertinent to crusted soils is that of Horton (Baver et al., 1972; Hillel, 1980; Morin and Benyamini, 1977):

$$i = i_c + (i_o - i_c)e^{-kt} \quad [1]$$

where i is the infiltration rate, i_c is the asymptotic final infiltration rate (FIR) reached when the time (t) (and the cumulative volume of water that had infiltrated) becomes large, i_o is the initial infiltration capacity of the soil, and k is a constant that determines how quickly i_o will decrease to i_c.

Several studies have been conducted on the effect of rainfall on the structure and hydraulic properties of soil crusts (Chen et al., 1980; Evans and Buol, 1968; McIntyre, 1958; Morin et al., 1981). McIntyre (1958) found that the crust consists of two distinct parts: an upper skin seal attributed to compaction by raindrop impact, and a "washed-in" region of decreased porosity attributed to the accumulation of particles. McIntyre (1958) measured thicknesses of 0.1 and 2 mm for the skin seal and the washed-in zone, respectively. The washed-in layer was formed only in easily dispersed soils. Chen et al. (1980) examined scanning electron micrographs of crusts of loessial soil and also found a thin skin seal about 0.1 mm in thickness that had formed at the uppermost layer of the soil. They did not, however, find an accumulation of fine particles in the 0.1- to 2.8-mm zone (the washed-in layer) as observed by McIntyre (1958).

Crust formation in soils exposed to the beating action of falling drops results from two mechanisms (Agassi et al., 1981; Kazman et al., 1983): (1) breakdown of soil aggregates caused by raindrop impact reduces the average pore size and compacts the surface layer, producing the thin skin seal at the soil surface; and (2) physicochemical dispersion of the soil clays, which migrate into the soil with

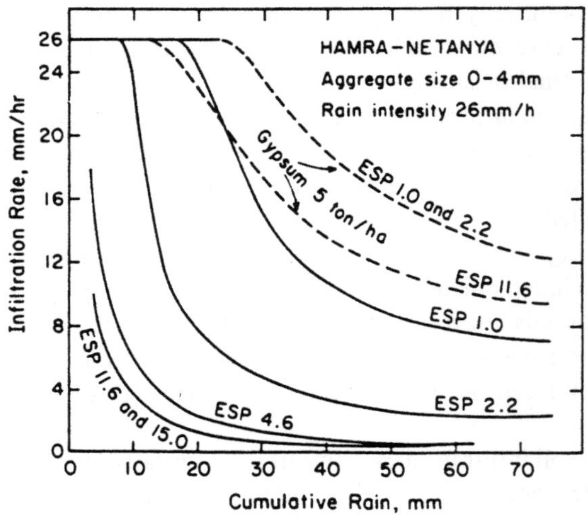

Figure 8. The effect of soil ESP and phosphogypsum on infiltration rate of a sandy loam (Netanya) soil (Kazman et al., Effect of low levels of exchangeable Na and applied phosphogypsum on the infiltration rate of various soils, *Soil Science*, 135, 184–192, © 1983).

the infiltrating water to clog the pores immediately beneath the surface (the washed-in zone).

A tool for studying the effect of soil sodicity, irrigation water salinity, and gypsum application on IR and crust formation is the rain simulator, such as those developed by Morin et al. (1967) and Miller (1987b). Soil samples are packed in small trays, which are placed in the simulator at a predetermined slope and subjected to simulated rain. The rain intensity and the drop impact energy can be controlled, and the water chemistry can be changed according to experimental plan. When rainwater is to be simulated, water of electrical conductivity of 0.01 dS m^{-1} is applied. The volume of the runoff and percolating water is measured and the IR is calculated.

2. Effect of Soil Sodicity and Electrolyte Concentration on IR

The effect of ESP on the IR and crust formation of four soils varying in texture, clay mineralogy, and $CaCO_3$ content was studied by Kazman et al. (1983) using distilled water in a rain simulator. In each of the four soils, IR was highly sensitive to low levels of ESP. The results for two soils, a noncalcareous sandy loam and a calcareous silty loam, are presented in Figures 8 and 9. An ESP value of 2.2 was enough to cause a large drop in the final IR of the sandy loam. Similar effects of ESP were observed on the silty loam soil. The amount of rain required to approach the final infiltration rate was also affected by ESP (Figures 8 and 9). As the ESP of the soil increased, the depth of rain required to reach the final IR decreased.

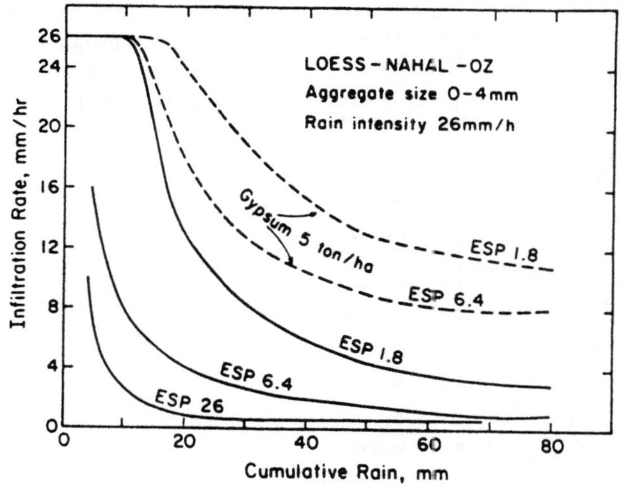

Figure 9. The effect of soil ESP and phosphogypsum on infiltration rate of a silt loam loess (Nahal-Oz) soil (Kazman et al., Effect of low levels of exchangeable Na and applied phosphogypsum on the infiltration rate of various soils, *Soil Science*, 135, 184–192, © 1983).

Because the raindrop impact energy was the same in all the experiments (Figures 8 and 9), the differences in IR curves for the various soil samples were the result of chemical dispersion caused by sodicity The IR is much more sensitive to ESP than is HC. The higher sensitivity of the soil surface to low ESP values is explained by three factors (Oster and Schroer, 1979; Kazman et al., 1983): (1) the mechanical impact of the raindrops, which enhances dispersion; (2) the absence of the soil matrix (sand particles), which slows clay movement; and (3) the low EC in the applied distilled water.

The mechanical and chemical mechanisms are complementary. The mechanical impact of the raindrops has two effects: (1) breakdown of the soil aggregates, followed by surface compaction and production of thin skin seals (McIntyre, 1958); and (2) stirring of the soil particles, which enhances their rate of chemical dispersion. Without the stirring process the rate of chemical dispersion is much slower. In HC measurements mechanical stirring of the soil is prevented, and a higher ESP is needed to cause dispersion.

Chemical dispersion may be prevented by spreading phosphogypsum (PG) (or another readily available electrolyte source) at the soil surface (Keren and Shainberg, 1981; and Figures 8 and 9). When 5 Mg ha^{-1} PG was applied to both soils, the IR drop and the steady-state IR were both affected. PG was effective also in soils of ESP 1.0, thus indicating that some chemical dispersion took place even at very low ESP values, and possibly even in Ca-saturated soils.

Keren and Shainberg (1981) found that PG was more effective in maintaining IR than mined gypsum because its rate of dissolution was much higher. The disso-

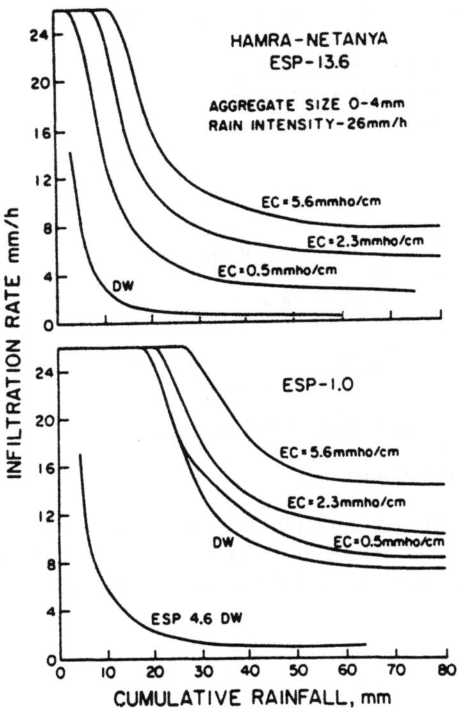

Figure 10. Effect of electrolyte concentration in rain simulation experiments on infiltration rate of a sandy loam soil (Agassi et al., 1981).

lution rate is important in IR studies because of the short contact time between rainwater and gypsum at the soil surface. The soil solution EC from surfaces treated with mined gypsum was apparently not sufficient to prevent dispersion and crust formation because of slow dissolution.

Chemical dispersion of the soil surface may also be prevented by "raining" with solutions of various EC using the rain simulator. Agassi et al. (1981) studied the effect of EC and soil sodicity on the IR of the two loamy soils described previously (Figure 10). They found that the IR curves of these soils increased markedly as the electrical conductivity of the applied water increased in the range between distilled water and 5.6 dS m^{-1}. Similar results were obtained by Oster and Schroer (1979). Agassi et al. (1981) concluded that crust formation was also a kinetic process. When solutions of high concentration are applied, the impact energy of the drops is the main force causing breakdown of the soil aggregates, and a compacted layer is formed at the soil surface. The rate at which this compacted layer is formed is relatively slow, and the final IR values are maintained at 8–15 mm h^{-1}. When low-salinity water is applied to the soil, even with low ESP chemical dispersion also occurs with the dispersed clay particles being washed into the soil with the infiltrating water and the pores immediately beneath the surface becoming clogged quite rapidly.

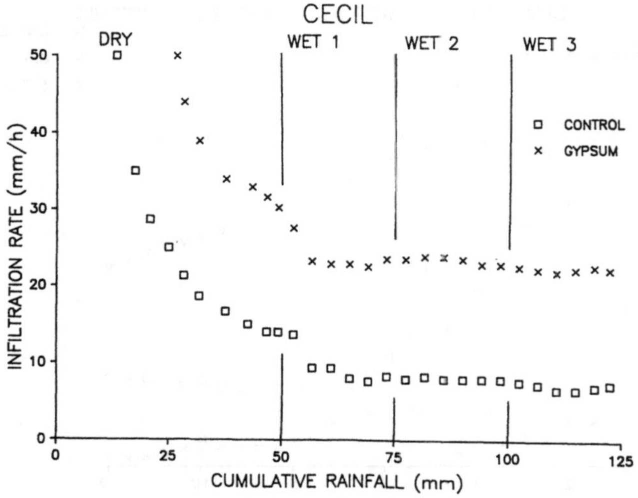

Figure 11. Infiltration rate of a Cecil sandy loam soil with (×) and without (□) surface-applied phosphogypsum (5 t ha^{-1}) during 125 mm simulated rainfall at 50 mm h^{-1} (Miller, 1987a).

The effect of PG on infiltration, runoff and soil loss of three Georgia Ultisols was studied by Miller (1987a). The effect of PG on the infiltration of Cecil soil exposed to simulated rain at 50 mm h^{-1} for 2.5 h [initial 1-h (dry) event and three subsequent 0.5-h (wet) events applied 24 h later] is presented in Figure 11. The Cecil soil is kaolinitic, with 8% clay and 17% silt, CEC of 3.15 cmol (+) kg^{-1}, and ESP of 1.2. In the absence of gypsum, IR stabilized at 10 mm h^{-1} with surface crusting and sand sorting apparent on the soil. Surface-applied PG maintained the final IR of the Cecil soil at 25 mm h^{-1}. The response of the kaolinitic Cecil soil to rain impact and PG application was quite similar to that of the Typic Rhodoxeralf (Netanya) soil from Israel (with ESP 1.0). It should be remembered, however, that the HC of the Rhodoxeralf, which contained montmorillonite, is more susceptible to sodic conditions (Figure 5) than that of the Cecil (Chiang et al., 1987). The fact that montmorillonitic soils from a semiarid region (Israel) and kaolinitic soils from a humid region (southeastern United States) disperse and crust in a similar manner when exposed to rain deserves further study. Nevertheless, the similarity in response to raindrop impact suggests that soil crusting and the correlation between crusting and soil dispersivity (Miller and Baharuddin, 1986) is quite a general phenomenon.

Highly weathered soils in the southeastern United States also suffer from dispersion-related degradation of physical properties with additions of Na, either from fertilizers or wastewater sources. The effect of surface application of $NaNO_3$, at fertilizer N rates, on infiltration, runoff and soil loss of a Greenville soil (clayey, kaolinitic, thermic, Rhodic Paleudult) was studied by Miller and

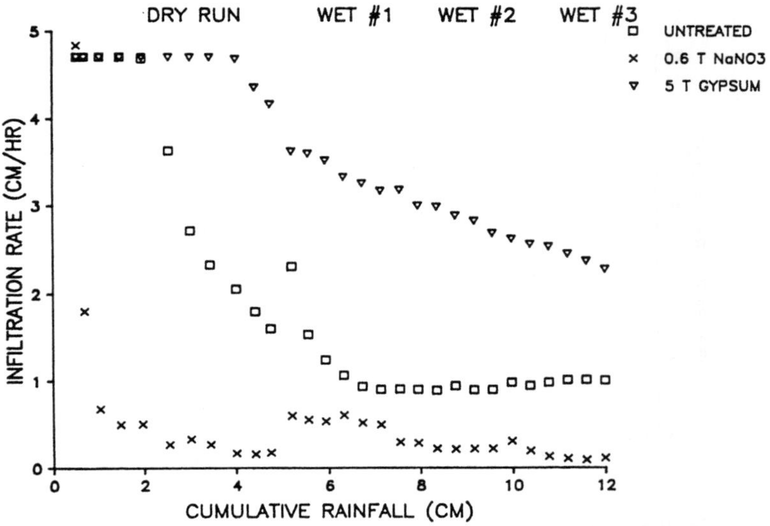

Figure 12. Infiltration rate of Greenville sandy clay loam soil treated with either 0.6 g ha^{-1} $NaNO_3$ or 5 t ha^{-1} $CaSO_4$, compared to untreated control, during 120 mm of simulated rainfall at 45 mm h^{-1}. (Reprinted with permission from W.P. Miller and J. Scifres, Effect of sodium nitrate and gypsum on infiltration and erosion of a highly weathered soil, *Soil Sci.*, 148, © by Williams & Wilkins, 1988.)

Scifres (1988) and is presented in Figure 12. Runoff from the untreated soil commenced after 2 cm of rainfall, and the steady-state IR stabilized at approximately 1 cm h^{-1}. Compared to other Piedmont soils (Miller, 1987a; and Figure 11), this soil maintained a reasonably high rate of water intake over the rainfall events, indicating a fair resistance to crusting. The $NaNO_3$ treatment (0.6 t ha^{-1}) resulted in nearly immediate surface sealing, as shown by the rapid decline in infiltration with a steady-state IR of only 2–3 mm h^{-1}. The application of PG had the opposite effect on infiltration: runoff was delayed until 45 mm of rain had accumulated, and the IR declined more slowly to a final value of 23 mm h^{-1} (Figure 12). Although the effect of exchangeable Na on the HC of kaolinitic soils is quite small (Chiang et al., 1987; Frenkel et al., 1978), the effects of low ESP values on clay dispersion, crusting, and hydraulic properties of crusts are significant. Significant changes in HC will take place only when spontaneous dispersion (Emerson, 1967) of aggregates takes place. Kaolinitic aggregates do not spontaneously disperse, and thus will not affect the soil HC. However, aggregates from kaolinitic soils can be mechanically dispersed, by the beating action of raindrops, to form a seal with low IR.

3. Management of Surface Crust with Phosphogypsum

Chemical dispersion can be prevented by spreading PG at the surface of sodic soils. The mechanism of PG interaction with the soil surface and the optimal rate of application to prevent surface seal formation is discussed next.

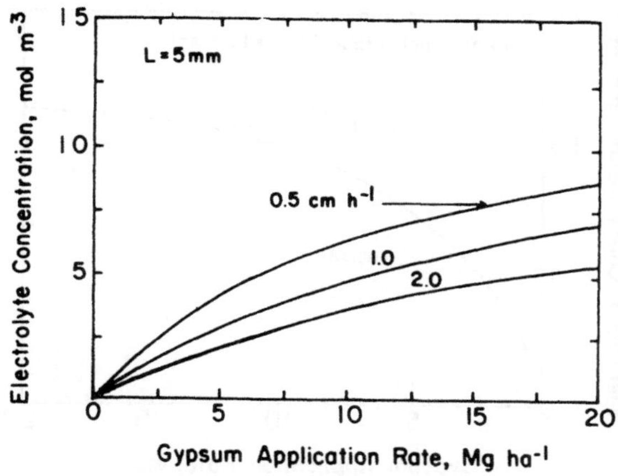

Figure 13. Average electrolyte concentration of the soil solution, if gypsum is mixed with the soil to depth of 5 mm, as a function of gypsum and water application rates (Oster, 1982).

a. Kinetics of Gypsum Dissolution

The assumption that gypsum dissolution was sufficiently rapid to maintain a saturated gypsum solution was reasonable in the reclamation models for sodic soils (see Section II.C.). However, equilibrium conditions probably do not apply to the dissolution of surface-applied gypsum. Here, the kinetics of dissolution of gypsum are limiting because of the shallow depth of the gypsum–soil layer and to the high water flux rates associated with the initial stages of infiltration.

Gypsum dissolution increases with increasing surface area of the gypsum particles, is controlled by film diffusion, and follows a first-order kinetic equation (Barton and Wilde, 1971; Keisling et al., 1978; Kemper et al., 1975; Keren and Shainberg, 1981; Oster, 1982). The EC of water flowing through a bed of gypsum particles was described by a transport equation that included terms to account for convection, diffusion, and dissolution kinetics (Keisling et al., 1978). This equation was used by Oster (1982) to illustrate the effects of gypsum, water application rates, and depth of mixing on EC. The effect of gypsum application and pore-water velocity on the average concentration of the soil solution is shown in Figure 13 for a mixed gypsum–soil layer 5 mm thick (Oster, 1982). For a given rate of water application, the average concentration increases with gypsum application rate because the amount of gypsum, and its associated surface area, increases per unit volume of soil. Increasing the water application rate decreases the contact time available for dissolution. Consequently, the EC decreases with increasing water application rate for a given gypsum application.

The preceding information is based on the dissolution rate of mined gypsum. According to Keren and Shainberg (1981), the rate of dissolution of PG can be 10 fold greater. The effect of such an increase on EC is shown in Figure 14. The concentrations for PG are about 3 times greater than for mined gypsum at applica-

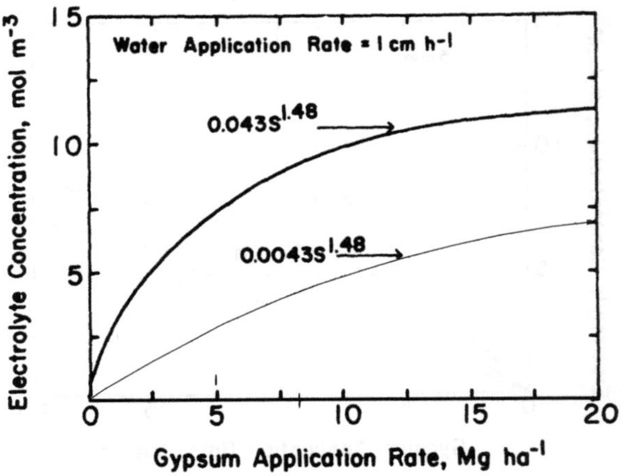

Figure 14. The effect of a 10-fold increase in dissolution rate of gypsum on the average electrolyte concentration of the soil solution; the depth of mixing is 5 mm and the water application rate is 1 cm h^{-1} (Oster, 1982).

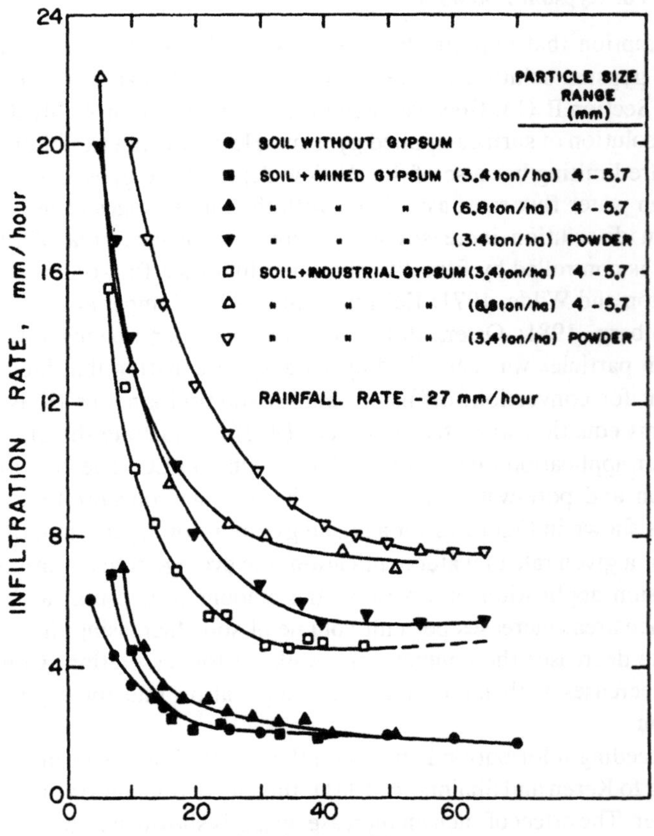

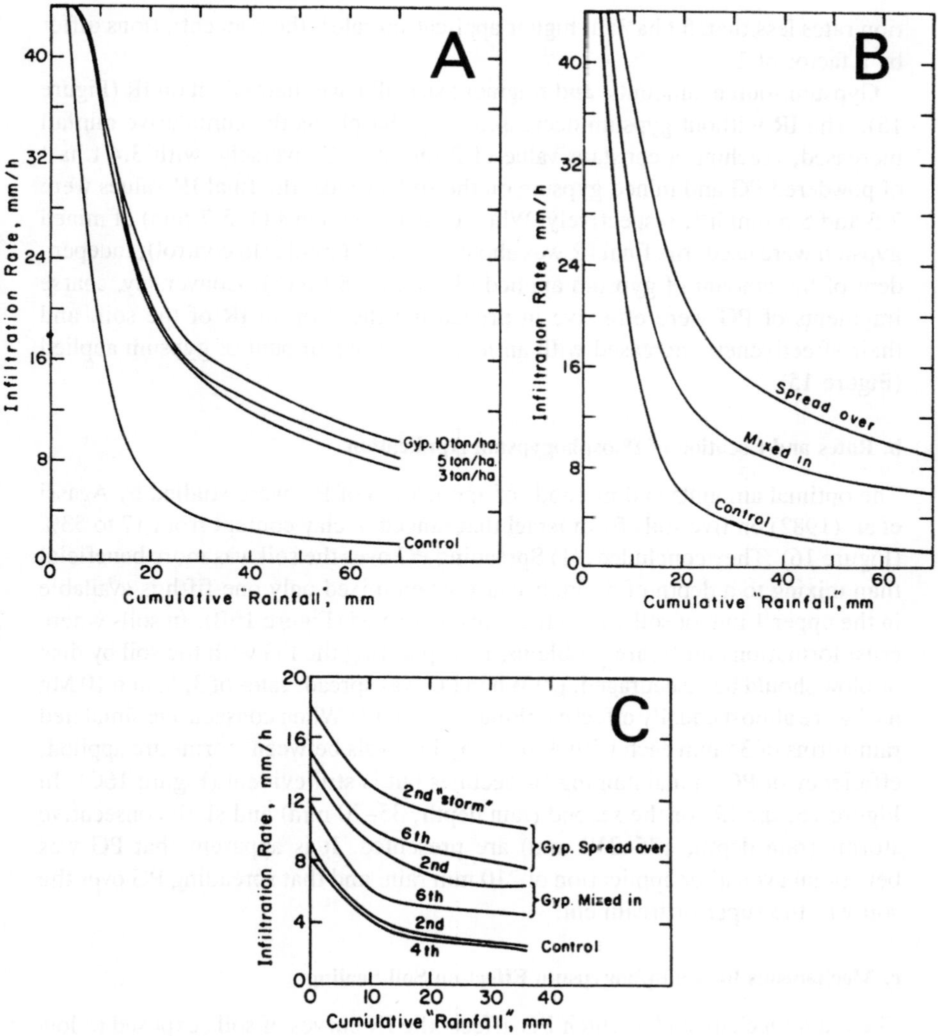

Figure 16. Infiltration rates of loess (Nahal-Oz) soil as a function of cumulative simulated rainfall: (A) The effects of quantities of applied gypsum; (B) the effects of method of application; (C) the effects of consecutive "rainfalls" and gypsum application (Agassi et al., 1982).

◄

Figure 15. The effect of rate and particle size of applied phosphogypsum and mined gypsum on infiltration rates of loess soil during 70 mm of applied simulated rainfall (Keren and Shainberg, 1981).

tion rates less than 5 t ha^{-1}; at higher application rates, the concentrations differ by a factor of 2.

Gypsum source, amounts, and fragment size all have their effect on IR (Figure 15). The IR without gypsum decreased very sharply as the cumulative rainfall increased, reaching a constant value of 2 mm h^{-1}. Conversely, with 3.4 t ha^{-1} of powdered PG and mined gypsum on the soil surface, the final IR values were 7.5 and 5.5 mm h^{-1}, respectively. When coarse fragments (4–5.7 mm) of mined gypsum were used, the final IR was about 2 mm h^{-1} (similar to control), independent of the amount of gypsum applied (3.4 and 6.8 t ha^{-1}). Conversely, coarse fragments of PG were effective in preventing the drop in IR of the soil, and their effectiveness increased with an increase in the amount of gypsum applied (Figure 15).

b. Rates and Location of Phosphogypsum Application

The optimal amounts and methods of application of PG were studied by Agassi et al. (1982) on five soils from Israel that ranged in clay content from 17 to 53% (Figure 16). They concluded: (1) Spreading PG over the soil was more beneficial than mixing to a depth of 5 mm because when mixed only one-fifth is available in the upper 1 mm of soil where the crust is formed (Figure 16B). In soils where crust formation and IR are problems, incorporating the PG with the soil by disk or plow should be discouraged. (2) When PG was spread, rates of 3, 5, and 10 Mg ha^{-1} were almost equally effective (Figure 16A). (3) When consecutive simulated rainstorms of 35 mm each with 3- to 5-day intervals between storms are applied, efficiency of PG in maintaining IR declines but is still evident (Figure 16C). In Figure 16, the IR for the second (rain depth, 35–70 mm) and sixth consecutive storms (rain depth, 175–210 mm) are presented. It is apparent that PG was beneficial even after application of 210 mm rain, and that spreading PG over the soil was the superior treatment.

c. Mechanisms for Phosphogypsum Effect on Soil Sealing

The main mechanism by which PG affects the IR curves of soils exposed to low electrolyte rain is by its dissolution and release of electrolytes into the soil solution. If this is the only mechanism, then applying a saturated solution of PG should give an IR curve similar to that of soil treated with PG and rained on with distilled water. In Figure 17, the IR of a loess soil with ESP values of 2.5 and 21.0, treated with 5 Mg ha^{-1} PG powder and rained on with distilled water (DW), are compared with the IR of the same soil (without PG) rained on with solution of 0.01 M $CaCl_2$ or saturated PG solution. Although the salinity of the percolating water was similar (=2 dS m^{-1}) in both treatments, the IR curve of the saline water (SW) was lower compared with the PG treatment. It is evident that the PG powder acts on the IR of the soils not only through its effect on the EC of the percolating water.

Phosphogypsum powder present at the soil surface may affect the crust properties and IR curves by two additional physical mechanisms: by interfering mechanically with the organization of the skin crust and thus disturbing the

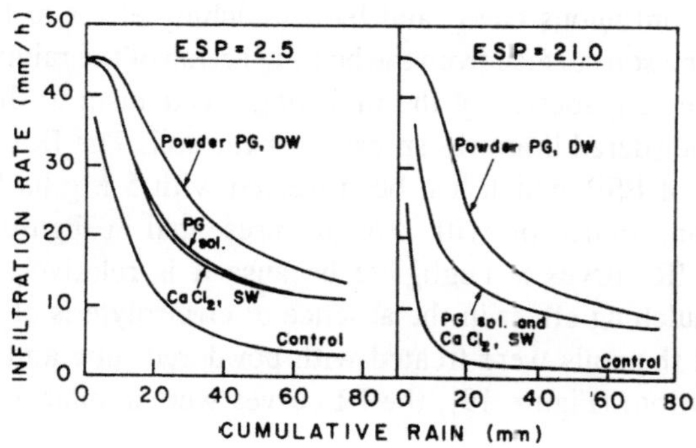

Figure 17. The effect of phosphogypsum (PG) powder and distilled water (DW) rain, compared with $CaCl_2$ solution (SW) and PG solution, on the IR of loess soil with high and low ESP (Agassi et al., 1986).

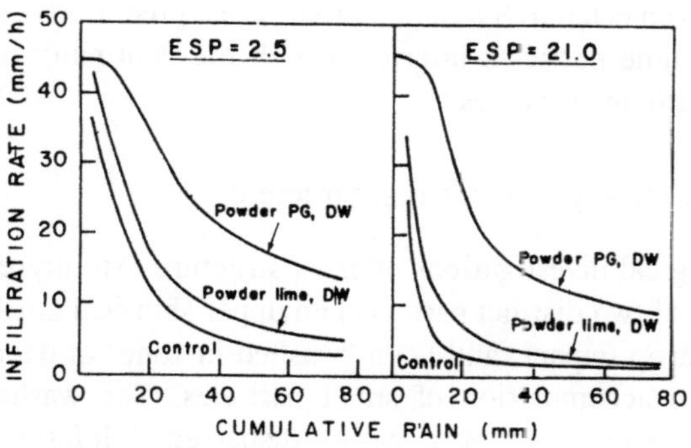

Figure 18. Effect of phosphogypsum (PG) powder and lime powder on the infiltration rate (DW, distilled water) of loess soil with high and low ESP (Agassi et al., 1986).

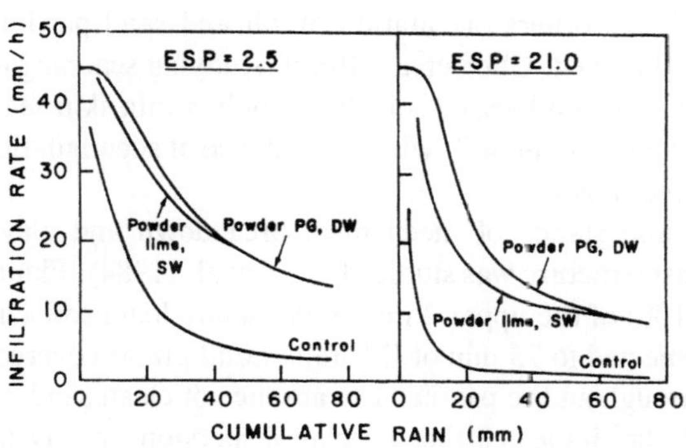

Figure 19. The effect of phosphogypsum (PG) powder rained on with distilled water (DW) and of lime powder rained on with salt water (SW, $CaCl_2$ solution) on the infiltration rate of a loess soil with high and low ESP (Agassi et al., 1986).

formation of a continuous curst; and by a mulching effect of the PG powder, which protects the soil surface from the beating action of the raindrops. To isolate and estimate the contribution of the mulching mechanism to the overall effect of PG powder, powdered lime was spread over the soil. The IR of the loess soils with two levels of ESP, which had been treated with 5 Mg ha^{-1} of either lime or PG powder and rained on with DW, are presented in Figure 18. The effect of lime on the IR curves is negligible because it is relatively insoluble, and therefore the mulching effect in the absence of electrolyte is of no importance. However, when the soils were treated with powdered lime and rained on with SW ($CaCl_2$ solution) (Figure 19), the IR curves were similar to those obtained with PG treatment.

When the EC of the percolating water was high, the mechanical effects of powdered lime and phosphogypsum on the IR were similar. The mechanical effect of powdered lime was less effective when DW was used because the "washed-in" crust predominated, and the lime particles at the surface were not effective in preventing the continuity of the crust. Conversely, when the powdered lime-treated soil was subjected to SW, a "skin seal" crust predominated at the soil surface where the lime particles interfered with the continuity of the crust and caused relatively high IR values.

4. Effect of Phosphogypsum on Crust Structure

Micromorphological investigations of crust structure (McIntyre, 1958) showed that it consisted of two distinct parts: (1) an upper skin seal attributable to compaction by raindrop impact; and (2) a "washed-in zone" of decreased porosity, attributed to the accumulation of small particles. The washed-in layer was formed only in soils that were easily dispersed (McIntyre, 1958). Low-magnification micrographs (×15) taken by Duley (1939) substantiate McIntyre's results. Higher resolution micrographs of soil crusts taken with a petrographic microscope, at magnifications up to ×100 (Evans and Buol, 1968), showed that in some soils particles in the layer immediately below the surface skin were oriented, whereas in others orientation of silt and sand particles occurred in deeper layers of the crust. Chen et al. (1980), studying scanning electron micrographs (SEM) of crust on loessial soil, found only a thin skin seal 0.1 mm thick at the uppermost layer of the soil, with no evidence of accumulation of fine particles in the washed-in zone.

The relative importance of mechanical breakdown and clay dispersion as expressed in crust structure was studied by Gal et al. (1984). Figure 20 shows the micrograph (×100) of the upper 2 mm of the sandy loam soil sample with ESP 1 after being subjected to 75 mm of DW rain. Sand grains covered with a skin of clay are seen throughout the profile. The absence of clean sand grains or a layer of accumulated clay indicates that no clay dispersion or clay movement took place in this profile. A thin (<0.1 mm) compacted layer of broken aggregates of low HC formed by raindrop impact sealed the soil surface, resulting in a final IR of 8 mm h^{-1}.

Figure 20. Scanning electron micrograph (SEM) of a sandy loam soil with low sodicity (ESP = 1). Original magnification, ×100 (Gal et al., 1984).

Figure 21 shows SEM micrographs of the Netanya sandy loam soil with ESP 11.6 after being rained on with DW. The surface layer (<0.25 mm) consists of large sand grains (Figure 21A). Low EC of the applied water and an ESP of 11.6 resulted in clay dispersion and stripping of the clay skin from the sand grains with some clay removal in the runoff water. However, some downward movement of clay and silt resulted in the formation of the washed-in layer (<0.75 mm thick) (Figure 21B).

Applying PG (5 Mg ha^{-1}) on the surface of the sandy loam soil with ESP 11.6 (Figure 22) prevented clay dispersion and movement without stripping of the clay skin from the sand grains and without clay accumulation in the washed-in zone. The decrease in IR resulted mainly from the breakdown and compaction of aggregates at the soil surface from the impact of raindrops.

Comparison of the SEM micrograph and IR of this soil (ESP 11.6) with and without gypsum indicates that clay accumulation in the washed-in zone occurs only when gypsum is absent. The effect of soil sodicity and PG on the IR and the micromorphology of a clay and a silty soil was studied by Gal et al. (1984) and Onofiok and Singer (1984), respectively. Crust permeability and IR of both soils were determined by the washed-in layer below the sand formed under conditions that are favorable for clay dispersion (high sodicity, distilled water, and mechanical impact of the raindrops), with no evidence of compacted layers ("skin seal") attributable to raindrop impact at the soil surface. Only "naked" sand grains were left on the soil surface, a condition typical for sodic soils exposed to rainstorms. These loosely held sand grains are eroded easily by the runoff water, exposing the washed-in layer at the soil surface. The appearance of a compacted layer at the soil surface under conditions favorable for clay dispersion is a secondary process following the erosion of the sandy layer.

The following conclusions may be drawn: (1) The impact of raindrops at the soil surface results in a compacted layer when the soil sodicity is very low (ESP <1.0) or in the presence of electrolytes from PG, which prevents clay dispersion, and the formation of a "washed-in" layer. The compacted layer, formed solely by the impact of raindrops, results in a final IR of 8–12 mm h^{-1}. (2) At ESP values greater than 3.0, clay dispersion takes place when the soil is exposed to distilled water rain. Because clay penetrates the soil and accumulates in the "washed-in" layer, very low final IR values result.

5. Effect of Phosphogypsum on Runoff in the Field

a. Measuring Infiltration in the Field

Because soil hydrological properties measured on disturbed samples are of little value in predicting field phenomena, in situ measurements are required. However, great spatial variability in infiltration and hydraulic properties of soils in the field is widely recognized (Russo and Bresler, 1981; Warrick and Nielsen, 1980) with coefficients of variation (CV) often exceeding 100%. This necessitated the application of geostatistical methods in which many measurements must be made even in a quite uniform field (Russo and Bresler, 1981; Vieira et al., 1981; Warrick and Nielsen, 1980). Conversely, variability in particle size

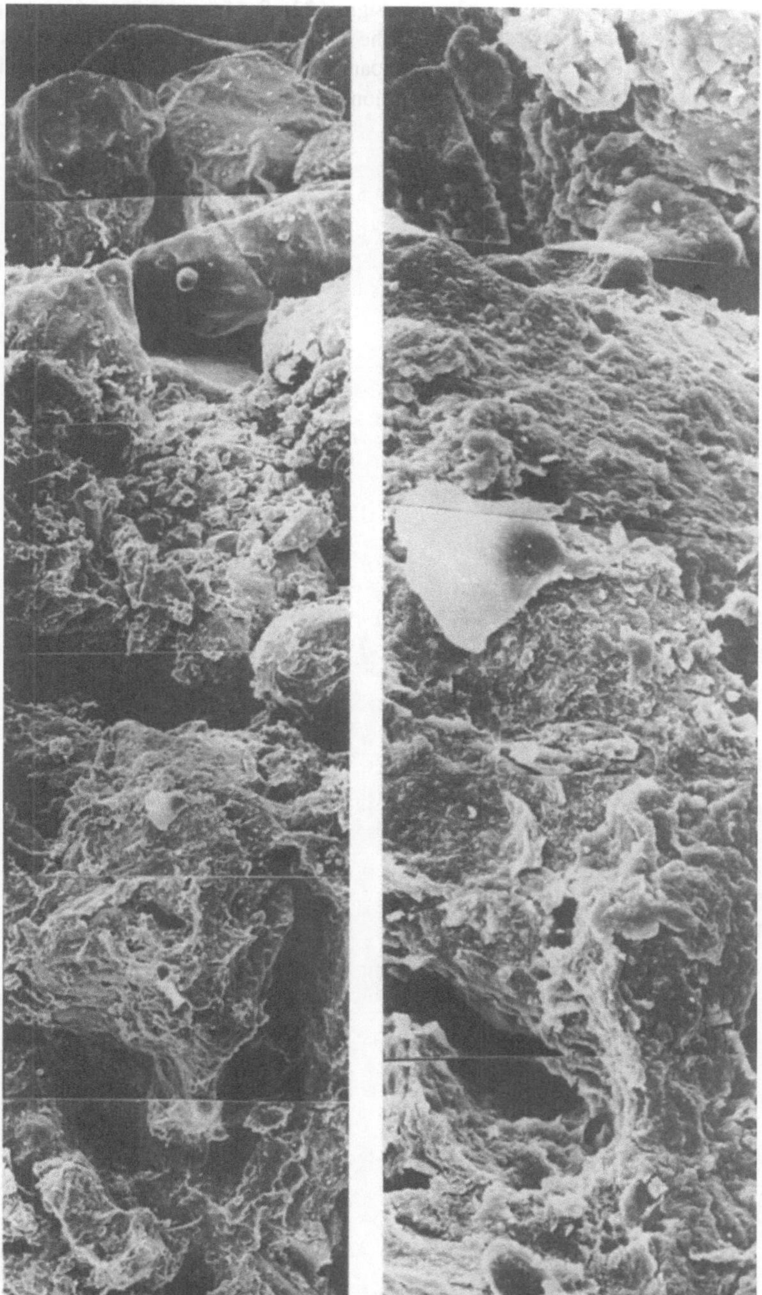

Figure 21. Scanning electron micrograph of the Netanya sandy loam soil with ESP = 11.6. (A) Crust profile, including surface, original magnification, ×300; (B) part of A, original magnification, ×1000. (From Gal et al., 1984.)

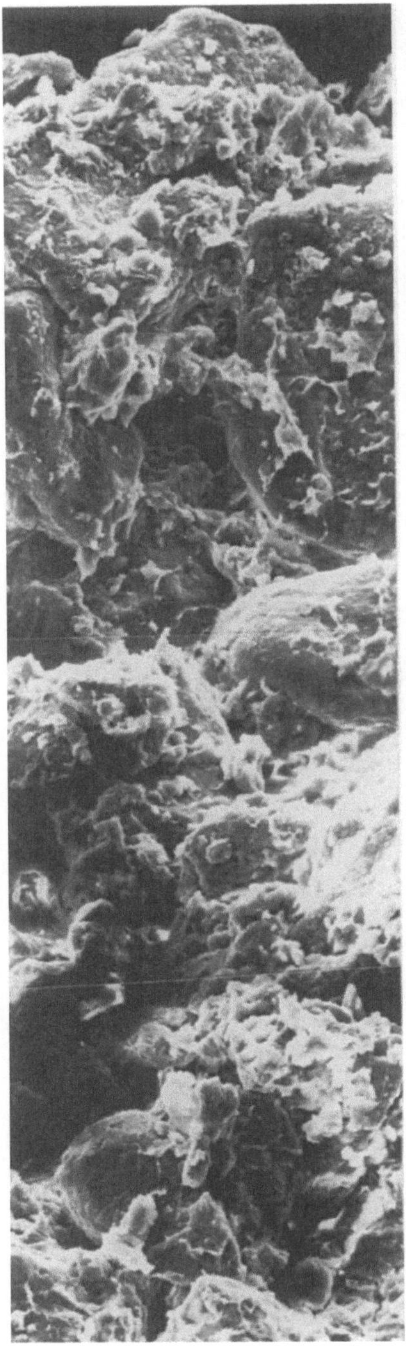

Figure 22. Scanning electron micrograph of the phosphogypsum-treated Netanya sandy loam with ESP = 11.6. Original magnification, ×300. (From Gal et al., 1984.)

and bulk density is intermediate (with CV values between 10% and 100%) and low (with CV values below 10%), respectively.

Various modifications of double-ring infiltrometers, with minimum soil surface disturbance, have been used to measure water intake in the field (Vieira et al., 1981), with the decrease in IR with time resulting mainly from the increased depth of the wetted zone. When the wetting front is deep enough, the matric potential gradient tends to zero and the steady-state IR approximates the saturated HC of the soil (Baver et al., 1972; Hillel, 1980).

Sprinkler infiltrometers that simulate rainfall in which water drops beat the soil surface and tend to disrupt the aggregates, compact the upper soil layer, and seal the pore space with fine particles forming a crust on drying have also been used to determine soil IR in the field. It is to be expected that the IR measured by sprinkler will be different from that measured by the double ring.

Ben Hur et al. (1987) studied the effect of surface seal formation on water intake and the variability and distribution of IR values measured under a portable field rainfall (raindrop impact conditions) and double-ring infiltrometers (flooded conditions). The mean steady-state IR values measured under flooded and sprinkler conditions were 57.8 mm h^{-1} (CV = 41.8%) and 8.6 mm h^{-1} (CV = 14.7%), respectively. This difference was caused by the formation of a soil surface seal from the impact of falling drops. The seal's HC was much lower and less variable than that of bulk soil. The low CV of steady-state IR obtained under raindrops suggests that the surface seal has a more uniform structure than that of the bulk soil underneath. This is probably because the upper layer of the soil is being mixed by cultivation. The low variability of IR in surface-sealed soil makes unnecessary large numbers of field measurements and the use of complicated geostatistical methods.

b. Effect of Gypsum on Infiltration and Runoff in Dryland Farming

Annual rainfall in semiarid regions is low in amount and uncertain in distribution, but, paradoxically, large amounts of water are lost to runoff. The amounts of runoff measured in the northern part of the Negev in Israel ranged from 30 to 50% of the rain in small (6-m^2) plots (Hillel, 1980). Surface runoff in cultivated fields is mainly the result of crust formation on the soil surface during rainstorms (Morin and Benyamini, 1977). Crust formation may be prevented by spreading PG at the soil surface.

The effect of PG application on the runoff/rain relation and yield of winter wheat under the complexity of field conditions on several soils was tested by Agassi et al. (1985b). Phosphogypsum at rates of 0, 5, and 10 t ha^{-1} was spread over the soil in runoff plots (5 × 25 m) following the seeding. In a few experiments, runoff was measured from plots with a wheat seedbed but no seeds and no PG (the "exposed" treatment). The difference in runoff between the control and exposed treatment was attributed to the effect of the wheat canopy on crust formation and runoff. In other experiments, the effects of slope length (5 and 25 m) on the runoff/rain ratio were studied. The yearly amounts of rainfall, runoff, erosion, and wheat yields in the runoff plots are given in Table 3. The precipitation

Table 3. Rainfall, runoff, erosion, and wheat yields in the runoff plots

Location (soil type)	Year	Rainfall (mm)	Treatment[a]	Runoff (mm)	Soil erosion (Mg ha^{-1})	Yield (Mg ha^{-1})
Dvir	1980/1981	229	Control	39.4	1.4	2.55
			PG, 5 Mg ha^{-1}	11.1	0.3	3.14
			PG, 10 Mg ha^{-1}	5.7	0.1	3.14
(loess)	1981/1982	159	Control	7.3	0.13	0.48
			PG, 5 Mg ha^{-1}	0.3	0.10	0.69
			Exposed (no wheat, no PG)	10.9	0.3	
Alumin	1980/1981	217	Control	23.0	–	0.98
			PG, 5 Mg ha^{-1}	12.2	–	1.22
			PG, 10 Mg ha^{-1}	12.5		1.28
(loess)	1981/1982	249	Control	23.4	0.5	0.43
			PG, 5 Mg ha^{-1}	3.6	0.1	0.88
			Exposed	28.0	0.4	–
	1982/1983	633	Control 25 m	124.3	1.54	3.10
			PG, 5 Mg ha^{-1}	34.1	0.60	3.10
			Control 5 m	212.0	4.87	–
Qedma	1980/1981	362	Control	50.6	11.2	1.65
			PG, 10 Mg ha^{-1}	23.2	3.4	2.00
(grumusol)						

From Agassi et al. (1985b).

[a] Control, wheat and no phosphogypsum. PG, 5 and 10 Mg ha^{-1}, indicate wheat field spread with 5 and 10 Mg ha^{-1} phosphogypsum, respectively. Exposed, wheat seedbed without wheat and PG.

in the winters of 1980–1981 and 1981–1982 was only 60–70% of the annual average, whereas that in the winter of 1982–1983 was higher than normal. The amounts of runoff for each storm during the three winters at two locations are presented in Figures 23 and 24.

In the winter of 1980–1981 at Dvir, the first rain after seeding occurred in mid-December, when 88 mm fell within 44 h (Figure 23). The runoff from this storm was 27.4, 4.8, and 2.7 mm for the control, 5, and 10 Mg ha^{-1} PG treatments, respectively, with successive rainstorms much lower in amount and intensity. Compared with the control (Table 3), 5 and 10 Mg ha^{-1} PG reduced the annual runoff and conserved 28 and 33 mm of rain, respectively. As a result, the wheat yield increased by 0.59 Mg ha^{-1}. Although the runoff from the 10 Mg ha^{-1} PG plots was 5 mm less than that from 5 Mg ha^{-1} plots, there was no measurable difference in wheat yield between the two treatments.

Phosphogypsum was more effective in reducing erosion than in reducing runoff (Table 3). Soil erosion was reduced by two mechanisms: (1) reducing runoff, and (2) stabilizing structure at the soil surface, a combination of which leads to low erosion.

In the winter of 1981–1982, treatment with 10 Mg ha^{-1} PG was replaced by the “exposed” treatment. As shown in Figure 23, until the rainstorm of March 8,

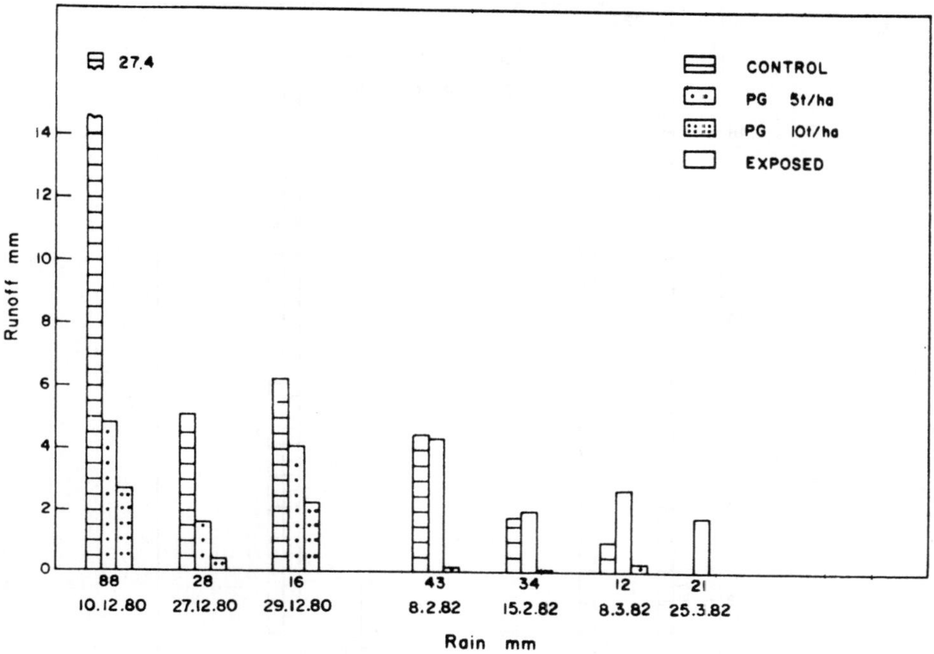

Figure 23. Runoff from control and phosphogypsum-treated plots at Dvir in the winters of 1980–1981 and 1981–1982 (Agassi et al., 1985b).

1982, there was little difference between the control and the exposed treatment in runoff. The accumulated amount of rainfall until that storm was 127 mm. Until the beginning of March, the wheat canopy in the control plots was not dense enough to prevent raindrop impact, thus crust formation and reduction in IR were similar in the control and exposed fields.

The amount of runoff from the March 8, 1982, rainstorm was higher in the exposed treatment than in the control. In March, air temperatures increased and the wheat canopy developed rapidly. The wheat canopy in the control treatment protected the soil surface from the impact of raindrops, and the IR of the soil remained higher than the IR of the unprotected "exposed" treatment. As a result of reducing runoff from 7.3 to 0.3 mm (Table 3), PG treatment increased wheat yield from 0.48 to 0.69 Mg ha^{-1}, which is probably not an economically significant increase.

The effect of plot length and PG on the amount of runoff from a loess soil with 2.5% slope is presented in Figure 24. There was a considerable difference in the amount of runoff between the 5- and 25-m-long plots in the first two rainstorms. Similar observations have been reported elsewhere (Lal, 1983). Observations during the rainstorm showed that much more runoff water was stored in the depressions on the soil surface of the longer plots. In order to clarify effects of this phenomenon, ditches were cut in the longer plots to connect the depressions to the outlet for runoff water. As a result, the amounts of runoff from the subsequent

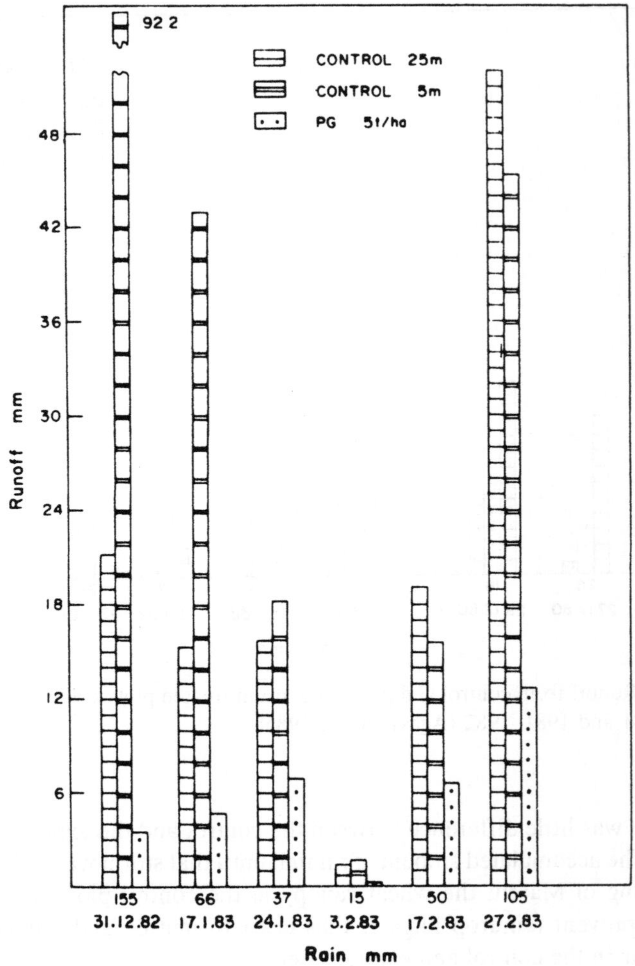

Figure 24. Runoff from control and phosphogypsum-treated plots at Alumin in the winter of 1982–1983 (Agassi et al., 1985b).

rainstorms were similar in the long and short plots. The high efficiency of PG was demonstrated again in that year. Even in the last rainstorm, after 528 mm of rain had fallen, PG reduced runoff by 76% compared with the control.

c. Effect of Gypsum on Rain Runoff in Irrigated Fields

Increasing use of saline water for irrigation in semiarid regions introduces Na to the exchange complex. During irrigation, high salt concentration prevents the adverse effects of Na, but when irrigation is replaced by rain, excess salts are leached, and the soil becomes almost impermeable as a result of crust formation. To prevent this condition during the rainy season, prior addition of PG is recom-

Table 4. Surface runoff, during rainstorms, from loess soil at two levels of ESP as affected by gypsum treatments

			Surface runoff (% of rainfall)				
			ESP 4.6		ESP 19.3		
Storm number	Amount of rainfall (mm)	Time between storms (days)	Control	Gypsum spread over	Control	Gypsum mixed in	Gypsum spread over
1	16	–	0	0	23.1	2.6	0
2	20	40	6.5	0	13.3	3.0	0.7
3	19	20	1.3	0.5	27.3	10.5	3.8
4	59	6	21.5	3.8	41.0	21.2	17.2
5	44	9	13.4	2.1	45.0	22.5	12.4
6	12	1	12.5	2.5	40.0	25.0	12.5
7	12	15	10.1	2.5	31.7	13.8	6.7
Total (mm)	182		22.8	3.8	64.2	30.1	18.7
Percent of annual rainfall			12.5	2.1	35.3	16.5	10.3

From Keren et al. (1983).

mended (Frenkel and Hadas, 1981; Keren et al., 1983). The effect of PG on water runoff from bare soil (Slope, 2–4%) with ESP 4.6 and 19.3 was studied by Keren et al., 1983 (Table 4). Even in a dry winter, with a total of only 182 mm of rain in seven storms, the total amount of runoff from the soils with ESP of 4.6 and 19 was 22.8 and 64.2 mm, respectively. In a wet winter (1982–1983), the recorded rainfall was 520 mm and the runoff from a 25-m-long plot of soil with ESP of 4.6 was 30% (157 mm). Phosphogypsum at 5 Mg ha^{-1} reduced the annual runoff from 22.8 to 3.8 mm and from 64.2 to 18.7 mm for the soils with ESP 4.6 and 19.3, respectively. High runoff from this calcareous loess soil with ESP 4.6, and the beneficial effect of PG, confirmed the hypothesis that even a low percentage of Na is enough to cause aggregate dispersion and crust formation. As the mechanical impact of the raindrops was constant, chemical dispersion (enhanced by sodicity and EC) plays a dominant role in affecting the IR.

E. Gypsum and Soil Conservation

1. Introduction

Soil erosion by water is a major worldwide problem that reduces soil thickness and consequently water and nutrient storage. Although soil lost from one field may be deposited on another, transport into a stream, river, or lake often results in pollution by sediments containing nutrients and pesticides. Thus erosion poses two distinct problems: a loss of fertility and soil depth in the agricultural field, and the addition of unwanted sediment in the aquatic environment.

Water erosion begins when a water drop strikes bare soil, detaching soil particles from the surface with subsequent transport downslope by raindrop splash or by overland flow (runoff). In soil erosion, raindrop detachment and overland flow are the two most important parts of the process in interrill soil areas.

Water flows over the soil surface when rainfall rate exceeds the soil IR. While many soils can initially absorb the water supplied by heavy storms, raindrop impact reduces IR by sealing the surface as a result of dispersion. Once sufficient water has accumulated on the soil surface, runoff commences. If the runoff is in thin sheets, sheet flow and sheet (or interrill) erosion result. As the water velocity increases in excess of 0.30 m s^{-1} (Ellison, 1947), turbulent flow causes rills to develop. As water becomes concentrated in rills (and gullies), faster movement detaches and transports soil particles more efficiently, causing rill erosion.

Soils resist water erosion differently, with some being more erodible. Several factors influence soil erodibility. Soils with stable aggregates are less erodible than others because they do not seal and runoff is limited; also, the stable aggregates are too large to be transported easily. Thus, soils low in exchangeable Na but high in clay, organic matter, and sesquioxide content are less erodible because of stable soil structure (Wischmeier and Smith, 1978). Conversely, soils high in silt have poor structure and are more erodible (Wischmeier and Mannering, 1969). Singer et al. (1982) found that soil erodibility increased significantly with the addition of as little as 2% Na to the exchange complex. Miller and Baharuddin (1986) found that for the kaolinitic soils of the southeastern United States a good correlation exists between soil erodibility and clay dispersion. Other studies have shown significant correlations between dispersion, runoff, and erosion in midwestern United States (Wischmeier and Mannering, 1969) and Hawaiian soils (El-Swaify and Dangler, 1977). It may be concluded that dispersive soils are more erodible than nondispersive soils, and clay dispersion may be an important predictor of soil erodibility in soils from humid to semiarid climates.

2. Effect of Phosphogypsum on Soil Loss on Gentle Slopes (<10%)

Surface-applied PG increases the EC of the rainwater, increases rain penetration, prevents soil dispersion and crust formation, and maintains larger aggregates at the soil surface. The reduced amount of runoff and the larger aggregates at the soil surface reduce erosion. The beneficial effect of PG in reducing erosion from kaolinitic and smectitic soils has been studied recently by Miller (1987a) and Warrington et al. (1988), respectively (Table 5). Because the amount, intensity, and impact energy of the rain are not the same for the two locations (Georgia, United States, and Israel), a quantitative comparison is impossible. In spite of the differences in mineralogy, all four soils developed crusts with final IR values less than 10 mm h^{-1}, with runoff accounting for 40–70% of the rain applied. Surface-applied PG increased the final IR of the four soils by 250% and roughly doubled the amount of rain that penetrated the soil.

The effect of PG on soil erosion is explained as follows. Three processes affect sediment concentration in the runoff water (Rose, 1985): (1) raindrop detach-

Table 5. Effect of PG on infiltration, runoff, and soil loss from soils with gentle slopes

Soil	Soil properties				Final IR ($mm\ h^{-1}$)		Runoff (%)		Soil loss ($kg\ ha^{-1}$)	
	Clay (%)	Silt (%)	CEC	pH	Control	PG	Control	PG	Control	PG
Cecil[a]	8	17	3.15	6.3	10	25	41.7	12.7	266	96
Wedowee[a]	17	17	4.07	5.0	5	17	72.1	31.7	1135	442
Worsham[a]	19	26	3.02	5.2	4	12	74.3	55.7	1315	732
Rhodoxeralf [b]	18	7	11.7	6.8	2.9	8.4	65.5	32.3	9350	4140

[a]Georgia, U.S. soils (Miller, 1987a).
[b]Israeli soil (Warrington et al., 1988).

ment of sediment from soil surface into overland flow; (2) sediment entrainment by overland flow (or runoff detachment) as rill and channel flow; and (3) sediment deposition, a continually occurring process owing to sediment settling out under gravity. This process is very sensitive to sediment size and fall velocity, which determine the resultant sediment concentration.

Detachment by rainfall is a major process at low slope angle (Walker et al., 1977). The effect of PG on rainfall detachment is relatively small and consists of two opposing processes. The formation of a seal with high shear strength may diminish rain detachment in the untreated soil, and the stabilization of small aggregates at the soil surface by the PG treatment will also diminish rain detachment because of the size of the stable aggregates. Thus it is assumed that the net effect is that rain detachment is similar in treated and untreated soils.

The capacity of runoff water to cause interrill erosion depends on its velocity and turbulence, which, in turn, depend on the slope angle, the depth of runoff water, and the roughness of the soil surface. Phosphogypsum treatment reduces the volume of runoff to roughly half that of the untreated soil, and increases the roughness of the surface and the tortuosity of the flow paths, thereby decreasing the velocity of overland flow. Phosphogypsum treatment also increases the EC in both the runoff and percolating water, which prevents aggregate dispersion (Agassi et al., 1981) and results in larger particles, which are less erodible. The relatively high EC in the runoff water (10 $cmol(+)\ l^{-1}$) flocculates and promotes deposition of clay particles. (Miller (1987a), analyzing the effect of PG on sediment particle sizes in the runoff water, found 15–30% clay and 60–75% silt in control treatments whereas PG-amended soils showed no clay-sized sediment in the runoff water. Phosphogypsum treatments therefore enhance sediment deposition processes, resulting in lower soil loss.

3. Effect of Steep Slope and PG on Soil Loss

The water erosion processes require a slope gradient to transport sediment. Level land is eroded slowly because the runoff velocity is low. As slope increases, runoff velocity and erosion also increase. Smith and Wischmeier (1957) deter-

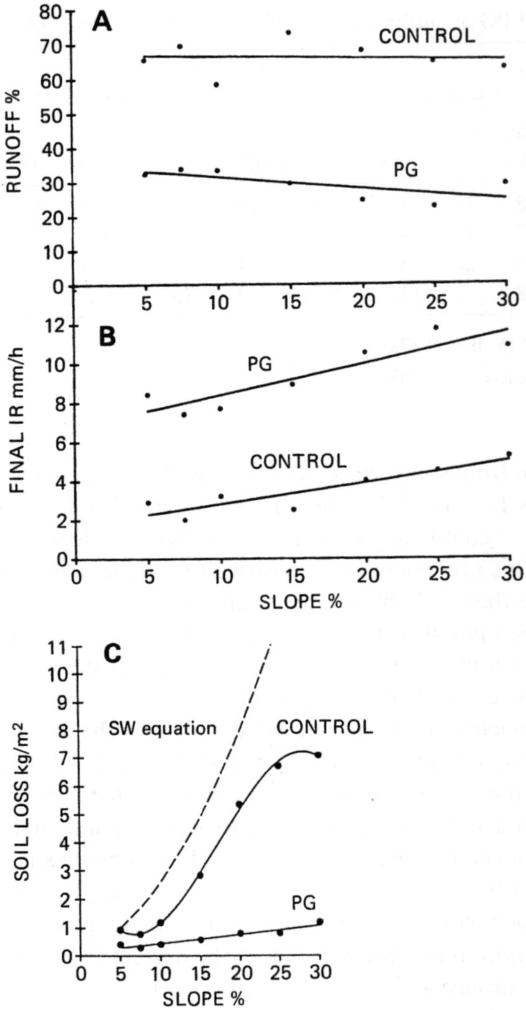

Figure 25. Effect of slope gradient on final infiltration rate (A), percent runoff (B), and soil loss (C) of a sandy loam soil under simulated rainfall (40 mm h^{-1} for 1.5 h) (Warrington et al., 1988).

mined empirically that soil loss was correlated with a parabolic description of the gradient. Normalizing their equation to a standard plot slope of 9% resulted in Equation [2]:

$$S = \frac{0.43 + 0.30s + 0.043s^2}{6.613} \quad [2]$$

where S = the slope gradient factor in the Universal Soil Loss Equation (USLE) (Wischmeier and Smith, 1978) and s = the gradient in percent. The effect of slope on IR and runoff in unstable soils, which tend to crust, is complicated (Poesen, 1987). Because the crust determines the low IR, with erosion of the crust, soil IR may increase with increase in slope. Poesen (1987) found that increasing slope decreased crust shear strength and increased IR because of reduction in the intensity of surface sealing. He concluded that as slope increases: (a) more splash and sheet erosion occurred, which removed the crust continuously; (b) the number of raindrop impacts per unit surface area decreased; (c) the normal component of drop impact force decreased; and (d) the occurrence of a thin film of water, which may cushion the compactive force of the raindrops, was reduced (Palmer, 1964).

The beneficial effect of PG in reducing erosion from steep slopes was studied by Warrington et al. (1988). As the slope angle increases (Figure 25A), final IR increases for both treated and untreated soils. These results for the untreated soil are similar to those obtained by Poesen (1987), and his explanations apply. The IR is determined by the surface seal formed by raindrops. As slope angle increases, more erosion of the seal occurs, and as the curst is removed, the IR increases. Erosion is not as pronounced in the PG-treated soil, and another explanation may be suggested. The sealing efficiency of the crust formed only by drop impact (when chemical dispersion is prevented by the presence of electrolytes released by PG dissolution) depends on the impact energy of the drops. As slope angle increases, the number of drops impacting a unit surface area and the normal component of drop impact force both decrease. As the impact energy of the drops decreases, a less developed seal is formed at the soil surface, and the IR increases (Agassi et al., 1985a).

The effect of slope on the percent of runoff is presented in Figure 25B. There was no obvious relationship between slope angle and the total amount of runoff from the untreated soil during the storm. The amount of runoff is affected by changes in the final IR, by the time to ponding and runoff, and by the rate of seal formation. In the case of the untreated soil, there was a tendency for percent runoff to decrease with increasing slope angle, while final IR increased. The interaction of these two relationships does not produce a simple trend in the cumulative infiltration, and changes in the amount of runoff during the storm are small. This is not the case when PG is added to the soil, because a relationship between runoff and slope angle is not apparent while final IR increases with slope angle. Thus a definite relationship (linear correlation coefficient; $r = .8$) between slope angle and cumulative infiltration exists with a decrease in runoff as the slope angle increases.

The effect of slope angle and PG treatment on soil loss is presented in Figure 25C. Increase in slope angle from 5 to 25% resulted in increased soil losses of 668% for the untreated and 190% for the PG-treated soils, respectively. The following observations should be noted: (a) the effect of slope angle on the measured soil loss is less than that predicted by Smith and Wischmeier's (1957) equation for the control soil; and (b) the rapid increase in soil erosion of untreated soil only occurs on slopes greater than 10%, with soil loss nearly constant between 25% and 30% slopes. Warrington et al. (1988) noted that this small change in erosion is probably an artifact caused by the walls of the flume. In the field, soil loss is expected to increase with increase in slope angle.

These observations were explained by the effect of crusting on surface shear strength and erosion. Soil compaction by raindrop impact increases the resistance of the soil to splash and sheet erosion, and the smoothness of the crust reduces turbulence in the runoff flow. Crust resistance to erosion was suggested as one reason why soil loss from all slopes was less than predicted by the Smith and Wischmeier equation. In addition, the protection afforded by the crust explained the resistance to erosion on the gentle slopes (<10%). At low slope gradients with low overland flow velocities, removal or detachment of sediment from the soil surface was mainly the result of rainfall detachment. As rainfall detachment is insensitive to slope, there is only a slight increase in erosion with increasing slope at slopes below a critical level. Walker et al. (1977) found that this critical slope, where rainfall is the only agent of sediment detachment and surface flow acts simply to transport and not generate sediment, is less than 3%. Because of the seal protection in the Rhodoxeralf, this slope increases to less than 10%. As the slope increases to more than 10%, the velocity of overland flow exceeded the critical value, rills started to form, erosion removed and weakened the crust, and soil loss increased at an accelerated rate.

The effect of PG on the microrelief of the soil exposed to rain at 5% and 30% slopes is shown in Figure 26. Following a rainstorm, the surface of the untreated soil at 5% slope is smooth. An untreated soil will initially form pits as a result of the impact of the raindrops (Hardy et al., 1983). As the storm proceeds, a crust is formed, and runoff and erosion take place. Because crust formation increases the resistance of the soil surface to further pitting, erosion removes loose soil particles, resulting in the development of a smooth surface. Soil treated with PG continues to pit throughout the storm, probably because the strength of the crust is not sufficient to resist the impact of the raindrops (Figure 26).

Increasing the slope angle increases the velocity of the water flowing over the soil surface. When the velocity and turbulence become large enough, rill formation takes place. On the 30% slope at the end of the storm, rills were as much as 0.75 cm deep and 2 cm wide on untreated soil, while PG-treated soil did not develop rills even on the 30% slope (Figure 26).

Phosphogypsum treatment, by releasing electrolytes to the soil surface and runoff solutions, decreased soil erosion on the steep slopes by (a) increasing the fraction of rain that penetrated the soil, thus decreasing the volume of runoff; (b) increasing the stability of the soil aggregates at the surface, with fewer soil particles being detached by rain impact and overland flow; (c) maintaining rough surfaces at the soil surface (compared with the smooth surface at the untreated soil), thus decreasing the velocity of the overland flow of runoff water; (d) enhancing the deposition of the clay fraction because of the high concentration of electrolytes in the runoff water. The beneficial effect of PG in preventing erosion from steep slopes may be useful in stabilizing natural or artificial steep slopes.

Figure 26. Appearance of soil surfaces of sandy loam soil after simulated rainfall (40 mm h^{-1} for 1.5 h) at 5% and 30% slopes (top and bottom photos, respectively), showing the effect of surface-applied phosphogypsum (right-hand photos) on crusting and rill development. (Photos by I. Shainberg.)

F. Gypsum and Water Penetration in Surface Irrigation

Slow water penetration, or slow infiltration, is a major problem in crop production on sandy loams to clays on more than 1 million ha of land in California (Oster and Singer, 1984). Similar problems of low water penetration were described elsewhere (Bashir et al., 1979; Kemper et al., 1975). Most of the land in California is irrigated by furrow or basin irrigation. Typically, water penetration in the beginning of the summer (beginning of the irrigation season, after a wet winter) is high, and the rate of water penetration decreases as the number of irrigations increases. Crust formation at the soil surface accounts for this low water permeability (Oster and Singer, 1984).

Crusting can decrease significantly the IR of water into soils. Two types of crusts have been recognized: structural and depositional. Structural crusts, formed when water drop impact breaks down soil aggregates and compacts the upper surface, have been discussed previously. Depositional crusts form when soil particles suspended in water are deposited on the soil surface as the water infiltrates. Sediments in running water originate from the erosion of soil materials, for example, when water flowing in furrows erodes the furrow sides during irrigation. The deposition of suspended material and formation of depositional crusts on the furrow bottom is repeated in each consecutive irrigation.

Only recently have the mode of formation of depositional crusts and their effect on water infiltration been studied. Shainberg and Singer (1985) produced depositional crusts in the laboratory and measured the effects of EC on water IR. They found that the HC of the surface crusts was two to three orders of magnitude lower than that of the bulk soil underneath when the electrical conductivity of the applied suspension was less than 0.3 dS m^{-1}. At EC values greater than 0.5 dS m^{-1}, the HC of the crust was only one order of magnitude lower than that of the bulk soil. Similarly, when PG was spread on the soil surface before low-EC irrigation water, high rates of water penetration were maintained (Table 6). The high susceptibility of the depositional crust to EC in the irrigation water was similar to that of structural crusts (Agassi et al., 1981). Whereas the HC of the soils studied was not affected by the EC in the applied suspension, the HC of the depositional crust was very sensitive to EC. Shainberg and Singer (1985) assumed that the dispersion–flocculation status of the sediments in the suspension determined the hydraulic properties of the depositional crust formed. Depositional crusts formed from flocculated particles had an open structure and high permeability. Conversely, when dispersed particles formed the depositional crust, the HC of the crust was several orders of magnitude lower than that of the bulk soil. Because flocculation value of smectites and illites with ESP less than 5.0 is 3 mmol(+) l^{-1} (Oster et al., 1980), it was assumed that the hydraulic properties of depositional crusts in equilibrium with more dilute solutions were low, whereas the permeabilities of crusts in equilibrium with more concentrated solutions were high.

Phosphogypsum maintains rapid water penetration by two mechanisms: (1) gypsum dissolves quickly and releases electrolytes that flocculate the soil parti-

Table 6. The hydraulic properties of the bulk soil and the depositional crust

Soil	Sand (%)	Silt (%)	Clay (%)	CEC [cmol(+) kg^{-1}]	ESP (%)	Base HC K_b ($m\ s^{-1} \times 10^{-6}$)	HC of the crust ($m\ s^{-1} \times 10^{-6}$) DW	0.002^a	0.005^a	PG
Wyman	69.7	21.6	8.7	9.0	4.8	5.3 ± 1.1	0.005	0.008	0.138	0.082
Wyo	34.4	48.4	17.2	20.0	0.8	2.2 ± 0.8	0.004	0.007	0.032	0.314

From Shainberg and Singer (1985).

[a] EC of percolating water ($S\ m^{-1}$); DW = deionized water; PG = PG-saturated water.

cles, and (2) gypsum releases Ca salts, which replace the exchangeable Na and lower the tendency of clays to disperse. Gypsum was more effective in the Wyo soil than in the Wyman (Table 6) because the high rate of water flow in the Wyman soil lowered the gypsum effectiveness compared with the low rate in the Wyo.

Phosphogypsum at the rate of 5 t ha^{-1} was effective in maintaining high water penetration rates even after three 130-mm suspension applications (Shainberg and Singer, 1985). The long-lasting effect of gypsum was assumed to result from two processes: (1) the concentration of gypsum in the percolating water was less than that in a saturated solution, and (2) during the drying periods between irrigations, solution from the soil profile evaporated at the surface. Some gypsum returned to the soil surface through capillary rise during the drying intervals.

The micromorphology of the depositional crusts prepared from suspensions with various EC values was observed with a petrographic light microscope and a scanning electron microscope by Southard et al. (1988). The density and morphology of the crusts were strongly influenced by EC. Crusts formed from soil material suspended in 0.002 *M* Cl solution had the highest bulk densities (1.95 Mg m^{-3}); surface application of PG produced the lowest density crusts (1.41 Mg m^{-3}). Crusts produced from suspension of soil material in DW and 0.005 *M* Cl had intermediate bulk densities. Suspensions of soil material in DW produced crusts with highly birefringent layers of clay oriented parallel to the soil surface, and suspended material penetrated at least 5 mm into the bulk soil. No birefringent, oriented clay was observed in crusts deposited from more concentrated solutions. Deposition and infiltration of the dispersed particles from clay suspensions in DW produced dense surficial crusts with oriented clay layers parallel to the surface and very low permeability. At high salt concentrations, clays exists as floccules, and not as individual platy particles, and settling of flocculated particles forms a more porous crust with a more random orientation of particles (not parallel to the soil surface).

G. Gypsum and Seedling Emergence

The formation of surface soil crusts that restrict water penetration may also have an adverse effect on seedling emergence, causing poor stands of crops and added expense of replanting. The occurrence of crusts of high strength that inhibit emergence has been widely documented (Cary and Evans, 1974), and is likely in many cases to be related to clay dispersion at the soil surface. While nondispersive soils may also form emergence-inhibiting surface layers because of aggregate breakdown and compaction under rainfall, the formation of washed-in layers of clay in dispersive soils appears to enhance crust strength and thereby often severely limit emergence.

Several observations recorded in the literature have suggested surface-applied gypsum may be effective in reducing crust strength and increasing emergence on dispersive soils. Gypsum decreased crust strength by a factor of two when surface applied to two sodic soils from India before a simulated rainfall treatment

(Chaudhri and Das, 1977). Gypsum increased cotton emergence on a Greenville soil from Georgia from 10% (untreated) to 50%, with a concomitant decrease in crust strength (Bennett et al., 1964). In Australia, Loveday (1974) showed a general relationship between crust strength, emergence, and hydraulic conductivity for a range of dispersive soils, and demonstrated increases of 100% in emergence with gypsum treatment. Further work by Shanmuganathan and Oades (1983b) has confirmed the positive effect of gypsum on emergence, and shown the importance of reducing clay dispersion in establishing good crop stands. These conclusions have been implemented in a recommendation for gypsum application on dispersive, hard-setting wheat soils in Australia to improve both emergence and water acceptance (Howell, 1987).

Most of the reported emergence responses to applied gypsum have been in controlled laboratory studies, with only limited field testing. Observations in the southeastern United States (W.P. Miller, unpublished data) have shown that rainfall conditions after planting are a crucial factor determining crust formation and the probability of a response to gypsum additions. Crusting on most soils is promoted by high-intensity rainfall immediately after planting, followed by a drying period that hardens the crust formed during the rainfall event. Emergence responses are likely to be obtained by gypsum application only in years when such conditions are prevalent.

H. Yield Responses and Economics

On sodic soils with very poor physical properties, the use of gypsum for soil reclamation results in a marked improvement in tilth and water acceptance, often improving crop production dramatically. On such soils the economics of gypsum use are unquestionably positive. On marginally sodic or nonsodic soils that are yet dispersive, gypsum usage may increase yields through better water infiltration and/or stand establishment. In low-rainfall areas or regions with erratic, intense summer thunderstorms, increases of 50–100% in cumulative infiltration may have a significant effect on crop yields via improved water availability to crops. Field data on such responses are scarce, however, except for Australia, where gypsum use on red-brown wheat soils (Alfisols) has been extensively studied. Here field trials during 1984–1987 have shown increases of 20% or more in wheat yields in 9 of 12 experiments, with some soils giving increases of 50% (Howell, 1987). Most of these increases are attributed directly to higher moisture supply from better infiltration of gypsum-amended soils. Yield responses are generally higher on conventionally tilled soils than on no-till (direct-seeded) areas, where surface sealing is lessened because of the presence of crop residue (Howell, 1987; Hamblin and Howell, 1988).

Despite the promise of yield increases, the economics of gypsum use are uncertain on these soils. Although both mined and byproduct gypsum are inexpensive at the source of production (as low as $5 per ton for phosphogypsum, to $80 per ton for some mined gypsums), transport costs of the large quantities that are required increase the total cost substantially. In the southeastern United States, cartage of phosphogypsum from the Florida stockpiles 500 km to central Georgia could cost up to $30 per ton, resulting in total costs of a 2.5 t ha^{-1} application of $100, including spreading costs. This estimate is similar to Howell's (1987) estimate for Australia. Given that applications may last only 1 or 2 years before the gypsum is leached out and reapplication is necessary, substantial yield increases will be required to make such an outlay economically viable. Hamblin and Howell (1988) are of the opinion that gypsum usage is not a long-term solution to crusting problems in Australia, largely because of the limited supply of gypsum and cost considerations, but that gypsum has a useful role to play in an integrated crop management system where it is used to improve aggregation during establishment of no-till and/or pasture crops.

Other considerations may also affect the cost-effectiveness of gypsum use. New regulations in the United States, limiting erosion and sediment production on cropland under the Food Security Act, may intensify interest in use of gypsum to control sediment loss from cropland. Gypsum also has a potential role in slope stabilization and physical property improvement on construction sites and urban lands where cost would be a lesser consideration, as suggested by Warrington et al. (1988). However, the greatest potential for economic use of gypsum may lie in crusting soils with acid subsoil horizons (Ultisols and Oxisols), where gypsum amendment not only improves water penetration but ameliorates subsoil limitations to root growth. Yield increases of agronomic crops on these soils are often dramatic and readily offset costs of the amendment, as is demonstrated in Section III.

III. Gypsum As an Ameliorant for Acid Subsoils

A. Introduction

The detrimental effects on plant growth of factors associated with soil acidity, particularly of high levels of exchangeable Al, are well known. Historically, however, soil scientists and agronomists have addressed the problem of soil acidity in the context of the plow layer, that zone readily accessible to amelioration by conventional liming and plowing procedures and that portion of the soil profile usually most acid in temperate, highly developed agriculture. Until little more than a decade ago, the effects of soil acidity in subsoil horizons had received little attention. Recently, there has been a growing awareness among agriculturists that in many parts of the world, particularly in vast areas of the tropics and subtropics, acid conditions in subsoil horizons pose a major limitation to successful production of many important crop species (Bouldin, 1979). Haynes (1984), in a comprehensive review dealing with lime and phosphate in soil–plant systems, has stressed the need to develop practical methods of ameliorating acid subsoils, which is given considerable urgency by the fact that such soils occur most widely in those parts of the world struggling to achieve self-sufficiency in food production.

Figure 27. Maize root development severely restricted by subsoil acidity. Dolomitic lime ($15\ t\ ha^{-1}$) had been plowed in 3 years previously and no physical restrictions existed. (Control plot in experiment 1, Table 8; photo by M. Farina.)

Toxic levels of Al, sometimes accompanied by deficient levels of Ca, may drastically restrict or even entirely prevent root exploitation of subsoil horizons (Figure 27). Crops so affected are unable to exploit moisture reserves below ameliorated surface horizons and are, consequently, highly susceptible to short-duration midsummer droughts. Wolf (1975), for example, has observed in the Cerrados of Central Brazil that maize wilts after only 6 days without rain, even during the wet season. Where the level of subsoil acidity and the frequency of midsummer drought is high, sustained economic production of many major crop species is not possible. Regional agricultural development is likely to be severely curtailed.

Reliable estimates of the areas globally affected by subsoil acidity are not available. However, calculations based on large-scale soil maps indicate that the areas involved are truly vast. Acid, low-base-status soils (Oxisols and Ultisols) constitute more than 40% of potentially arable tropical soils (Sanchez, 1977), an area which has been estimated as being in excess of 800 million ha (Orvedal and Ackerson, 1972). In addition, there are reports in the literature of severe subsoil acidity occurring in subtropical and temperate latitudes (Adams and Moore, 1983; Doss et al., 1979; Farina and Channon, 1988a; McKenzie and Nyborg, 1984; Simpson et al., 1979; Summer et al., 1986), and intensive cropping accompanied by high N use may lead to subsoil acidification of soils that are not naturally prohibitively acidic (Abruna et al., 1964; Adams, 1984; Adeoye and Singh, 1985; Pearson et al., 1962).

Some examples of soils with subsoil acidity problems are given in Table 7. Significantly, those from South Africa and the United States originate from localities well outside the tropics. The Balmoral soil typifies subsoil acidity in soils developed under lower rainfall regimes (~800 mm per annum) than those more commonly associated with the problem. A zone of high acidity, often less than 0.30 m wide, occurs in the immediate subsoil and deeper horizons, which although not physically hostile to root development, remains inaccessible because of the chemical barrier imposed by Al. Soils acidified by intensive cropping usually exhibit a similar pattern. The Brazilian soil, developed under very intensive weathering conditions, is typical of the Oxisols that dominate tropical regions (Bouldin, 1979; Sanchez, 1977).

Because of the low mobility of lime, or at least of its alkaline component, subsoil acidity in most soils cannot be rectified by conventional incorporation of lime in the plow layer (Brown and Munsell, 1938; Farina and Channon, 1988b; Pearson et al., 1973; Rechcigl et al., 1985; Sumner et al., 1986). Figure 27 illustrates this well. While vertical movement of Ca and Mg, mainly as nitrates, may be appreciable in spite of the negative charge created by pH increases in variable-charge soils (Pearson et al., 1962; Ritchey et al., 1980, 1982), OH^- and HCO_3^- ions are consumed by acidity in the topsoil unless massive, uneconomic levels of lime are applied. Consequently, where Al toxicity rather than Ca deficiency limits root growth, specialized management strategies are required to overcome the problem.

Strategies available include the use of equipment that will incorporate lime into the subsoil (Anderson and Hendrick, 1983; Doss et al., 1979; Farina and Channon, 1988a; Gonzalez-Erico et al., 1979; Hammel et al., 1985; Kauffman and Gardner, 1978; Pearson et al., 1973; Perez-Escolar and Lugo-Lopez, 1978; Sumner et al., 1986), the use of Al-tolerant crops or genotypes (Bouton and Sumner, 1983; Foy, 1974; Ponnamperuma, 1982; Sanchez and Salinas, 1981), and surface incorporation of gypsum (Farina and Channon, 1988b; Hammel et al., 1985; Pavan et al., 1982, 1984; Reeve and Sumner, 1972; Ritchey et al., 1980; Sumner et al., 1986).

All these options are beset with problems, and it is probably true to say that in many parts of the developing world none are yet truly viable propositions. Mechanical profile modification, while potentially effective economically (Farina and Channon, 1988a), has power requirements beyond the reach of many farmers. The second option, probably that with the greatest long-term potential on soils with only marginally toxic subsoils, is limited in the short term by the time span required to develop Al-tolerant genotypes, and by the fact that it is often not economically feasible to change from Al-sensitive grain crops to tolerant pasture species. To date, only Brazil has made significant use of this option. The third alternative, surface incorporation of gypsum, is limited to areas where gypsum is available at reasonable cost, either mined from natural deposits or as a by-product of phosphoric or other acid production. In the United States, Brazil,

Table 7. Chemical characteristics of some acid soils

Soil	Depth (m)	Organic carbon (%)	Clay (%)	Exchangeable cations ($cmol_c\ kg^{-1}$) Ca	Mg	K	Al + H	Acid saturation (%)	pH_{salt}	pH_{water}	Source
Normandien clay loam	0–0.15	1.5	41	0.65	0.30	0.50	2.06	59	3.91	4.50	Farina and Channon
(Plinthic Paleudult)	0.15–0.30	1.2	41	0.48	0.22	0.29	1.83	65	4.00	4.55	(1988a)
South Africa	0.30–0.45	0.8	46	0.26	0.11	0.16	1.78	77	4.08	4.46	
	0.45–0.60	0.5	49	0.26	0.12	0.14	1.78	74	4.06	4.40	
	0.60–0.75	0.4	47	0.28	0.18	0.12	2.13	79	4.02	4.27	
	0.75–0.90	0.3	46	0.32	0.41	0.11	2.48	75	3.97	4.35	
Balmoral clay loam	0–0.15	3.9	61	2.05	0.69	0.47	2.23	41	4.00	–	Farina (unpublished
(Typic Haplorthox)	0.15–0.30	–	–	0.75	0.35	0.26	2.66	66	4.05	–	data)
South Africa	0.30–0.45	2.4	67	0.64	0.27	0.18	1.65	61	4.18	–	
	0.45–0.60	–	–	0.92	0.40	0.16	1.18	44	4.28	–	
	0.60–0.75	–	–	1.38	1.10	0.09	0.25	9	4.49	–	
	0.75–0.90	–	–	1.21	1.44	0.09	0.35	11	4.47	–	
Appling coarse sandy	0–0.15	0.8	10	1.12	0.11	0.08	0.70	34	4.40	5.20	Sumner et al.
loam (Typic Hapludult)	0.15–0.30	0.6	21	0.70	0.22	0.06	1.48	60	3.92	4.86	(1986)
Georgia	0.30–0.45	0.3	33	0.68	0.33	0.06	1.63	60	3.94	4.88	
	0.45–0.60	0.2	46	0.97	0.22	0.05	1.70	57	3.95	4.76	
	0.60–0.75	0.2	26	0.65	0.19	0.05	1.96	68	3.93	4.80	
	0.75–0.90	0.1	21	0.44	0.10	0.04	2.21	79	3.96	4.89	
	0.90–1.05	0.1	16	0.29	0.08	0.03	2.40	85	4.01	4.85	
Latossol vermelho-exuro	0–0.15	–	–	0.10	–	–	–	65	–	4.60	Souza and Ritchey
(Oxisol)	0.15–0.30	–	–	0.09	–	–	–	60	–	4.60	(1986)
Brazil	0.30–0.45	–	–	0.08	–	–	–	61	–	4.25	
	0.45–0.60	–	–	0.09	–	–	–	60	–	4.23	
	0.60–0.75	–	–	0.08	–	–	–	55	–	4.20	
	0.75–0.90	–	–	0.08	–	–	–	50	–	4.90	
	0.90–1.05	–	–	0.08	–	–	–	35	–	4.90	

South Africa, and probably many other parts of the world affected by subsoil acidity, such opportunity exists. Dependent on the applicability of the alternative strategies mentioned, surface incorporation of gypsum may prove to be the most viable approach to the problem.

To date, research involving the use of gypsum as a subsoil ameliorant has been largely restricted to the United States, Brazil, and South Africa. Research programs currently being pursued in all three countries have their origins in findings originally published by Sumner (1970) and Reeve and Sumner (1972). Brazilian scientists in the Cerrados of Central Brazil were the first to pursue this work further (Ritchey et al., 1980), and rapid progress has been made during the past decade (Anonymous, 1986). Subsoil acidity is a particularly severe problem in Brazil because of the immense areas over which it occurs and the relatively much smaller areas of unaffected soil (Dematte, 1981; Goedert, 1983). The very high annual precipitation, 1,000 mm to 1,800 mm on the Cerrados (Goedert, 1983) and greater than 2,000 mm in much of the remainder of the country, has resulted in the development of exceptionally highly leached and weathered soils (Table 7) and subsoil conditions extremely hostile to root development. In addition, short midsummer droughts known locally as "veronicos" are not uncommon. According to Wolf (1975), central Brazil experiences 10 days or more without rain during the summer wet season 1 year in 2, and 2 weeks or more without rain 1 year in 5. Because maize has been observed to wilt after only 6 days without rain (Wolf, 1975), the need to improve subsoil conditions is clearly a national priority of the highest order.

Intensive research into the use of gypsum as a subsoil ameliorant was initiated in the United States in 1980 (Hammel et al., 1985; Sumner et al., 1985). In the southeastern United States, many soils typically have subsoils sufficiently acid to subtend substantial levels of soluble Al and Mn with very low levels of Ca and other nutrients (Table 7). These old, highly weathered soils (Ultisols) have developed under annual rainfall exceeding 1,200 mm, and despite the relatively wet conditions, midsummer droughts of 1–4 weeks duration, during which evaporation far outstrips precipitation, occur frequently. A common characteristic of these soils is the presence of well developed Bt-horizons, which can often develop physical hardpans at the Ap–B-horizon interface as a result of mechanical tillage operations. Thus, such soils often have both chemical and physical pans, which together limit root penetration. The final manifestation of these chemical and physical pans on plant performance is to be found in crop sensitivity to drought stress and resultant typically unstable yield levels. It has recently been clearly demonstrated that yields of many crops can be as much as doubled provided subsoil acidity is ameliorated by deep lime incorporated (Hammel et al., 1985; Sumner et al., 1986). Subsoil amelioration is regarded as being critical to the optimization of agricultural development in large parts of the area concerned.

In South Africa, the original findings of Sumner (1970) and Reeve and Sumner (1972) were only further pursued more than a full decade later (Farina and Channon, 1988b). In this country, the problem of subsoil acidity, which occurs extensively in the most favorable rainfall (800–1000 mm) areas, is exacerbated by the fact that although highly weathered dystrophic soils (Table 7) only make up approximately 15% of the area available for dryland cropping, the potential for vertical expansion in other areas is limited by low, very erratic rainfall. These highly weathered soils will, thus, have a major role to play in feeding a population expected to double during the course of the next two decades. Here too the negative effects of subsoil acidity are further aggravated by the high frequency of midsummer droughts. Typically, midsummer dry periods lasting 10 days or more occur 2 years in 3, and maize, the staple food in much of southern Africa, is extremely sensitive to moisture stress during this period. It is not uncommon for promising crops to be destroyed by droughts of little longer than a week, even when seasonal rainfall exceeds 600 mm. Development of economically viable methods of increasing subsoil root proliferation is seen as an urgent research need, and encouraging progress in this direction has recently been made (Farina and Channon, 1988a,b).

It seems probable that similar circumstances to those existing in the aforementioned three localities occur elsewhere, and it is the intention here to present data justifying the use of gypsum as a subsoil ameliorant. The data discussed have their origins in the American, Brazilian, and South African situations outlined, and represent the bulk of information currently available on the subject. While many questions remain to be answered, an effort is made to formulate unifying principles and to explain possible points of divergence.

B. Crop Response

1. Yield

Since the initial studies by Sumner (1970) and Reeve and Sumner (1972), extensive field experimentation in South Africa, Brazil, and the southeastern United States has documented unequivocally the beneficial effects of gypsum on soils with strongly acidic subsoils. There is also a substantial volume of research on the beneficial effects of gypsum under greenhouse conditions, but this has not been emphasized here because of the uncertain applicability of greenhouse data to field conditions. To illustrate the magnitude and nature of gypsum response in the field, a selection of results is presented in Table 8. Basically, two types of gypsum were used in these experiments; gypsum mined from natural deposits and phosphogypsum (PG) produced as a by-product in phosphoric acid production.

In all the experiments that have been conducted, it is clear from the cumulative responses obtained that very appreciable yield increases of maize, soybeans, coffee, wheat, rice, beans, peaches, apples, cotton, and alfalfa can be sustained. Despite the relatively high gypsum rates used in many of the trials, such responses would have covered the cost of gypsum within 3–4 years. Distance from source and transportation costs, however, are naturally critical considerations in this regard. Soil data (Section III.C) indicate that the beneficial effects of gypsum application (10 Mg ha^{-1}) are likely to persist well in excess of 6 years on the strongly acidic soils studied. In the case of the South African and

Table 8. Crop responses to gypsum application

Origin	Crop	Soil texture	Soil order	Gypsum rate (Mg ha^{-1})	Gypsum source	Year
1. South Africa	Maize	Clay loam	Ultisol	5	Mined 86%	1
				5	$CaSO_4 \cdot 2H_2O$	2
				0		3
				0		4
				0		5
2. South Africa	Maize	Clay loam	Ultisol	1	Mined 85%	1
				2	$CaSO_4 \cdot 2H_2O$	
				4		
				8		
				16		
				0		2
				0		
				0		
				0		
				0		
3. South Africa	Maize	Clay loam	Ultisol	2.5	Mined or	1
				5.0	Phosphogypsum	
				0		2
				0		
				0		3
				0		
4. Georgia	Soybean	Sandy loam	Ultisol	35	Mined	1
				0		2
	Maize			0		3

Table 8. *(Continued)*

Yield						
− Gypsum	(Mg ha⁻¹)	+ Gypsum	Change (%)		Remarks	Source
6.45	Silage	7.10	+10	NS	S nutritional effects eliminated. All plots received 15 $Mg\ ha^{-1}$ dolomitic lime. No grain in first year due to severe drought. Moisture stress in second and fifth year. No visible moisture stress in third and fourth seasons	Farina and Channon (1986b); Farina (unpublished data)
4.77	Grain	5.81	+22	**		
7.80	Grain	9.05	+16	**		
5.90	Grain	7.05	+20	**		
2.66	Grain	3.82	+44	**		
4.96	Grain	4.83	−3	NS	S nutritional effects eliminated. Gypsum applied at four lime rates (0, 3, 6, 12 $Mg\ ha^{-1}$). No interaction with lime evident.	Farina (unpublished data)
4.96	Grain	4.78	−4	NS		
4.96	Grain	5.07	+2	NS		
4.96	Grain	5.12	+3	NS		
4.96	Grain	5.11	+3	NS		
11.27	Silage	11.22	+4	NS	No interaction with lime evident. No grain due to severe heat stress at flowering. No response to gypsum rates of 1 and 2 $Mg\ ha^{-1}$ (see soil data, Table 15)	
11.27	Silage	11.38	+1	NS		
11.27	Silage	12.09	+7	*		
11.27	Silage	12.56	+11	**		
11.27	Silage	12.27	+9	**		
5.38	Grain	5.75	+7	NS	S nutritional effects eliminated. Topsoil Al eliminated in all plots. Response due to P in phosphogypsum	Farina and Fey (unpublished data)
5.38	Grain	6.01	+12			
3.22	Grain	3.73	+16	*	Significant response to both gypsum sources. No P effect evident	
3.22	Grain	3.97	+23	**		
10.19	Silage	10.52	+3	NS	No grain due to heat stress at flowering. No overall response to gypsum	
10.19	Silage	10.50	+3	NS		
0.94	Grain	1.22	+30	NS	S nutritional effects eliminated. Topsoil on all plots limed and fertilized in accordance with soil test recommendations	Hammel et al. (1985)
1.28	Grain	1.61	+26	*		
26.20	Silage	35.40	+35	**		

Table 8. *(Continued)*

Origin	Crop	Soil texture	Soil order	Gypsum rate (Mg ha^{-1})	Gypsum source	Year
5. Georgia	Alfalfa	Loam	Ultisol	10	Mined	1
				0		2
				0		3
				0		4
				0		5
				0		6
6. Georgia	Alfalfa	Loamy sand	Ultisol	2	Phosphogypsum	1
				5		
				10		
				0		2
				0		
				0		
7. Georgia	Alfalfa	Loam	Ultisol	10	Phosphogypsum	1
	Sericia			10		
8. Georgia	Peaches	Loam	Ultisol	10	Mined	1
				0		2
				0		3
				0		4
9. Georgia	Alfalfa	Loam	Ultisol	10	Phosphogypsum	1
				10	Flue gas gypsum	
				10	Flue gas gypsum + fly ash	
				10	Mined gypsum	
10. Georgia	Cotton	Sandy loam	Ultisol	10	Phosphogypsum	1
				0		2
				0		3

Table 8. *(Continued)*

Yield						
− Gypsum (Mg ha⁻¹)		+ Gypsum	Change (%)		Remarks	Source
9.73	Hay	10.36	+6	NS	S nutritional effects eliminated. Topsoil on all plots limed and fertilized in accordance with soil test recommendations. Years 4 and 5 were particularly dry	Sumner et al. (1986); Sumner (unpublished data)
7.18	Hay	9.08	+26	**		
10.67	Hay	13.59	+27	*		
3.29	Hay	7.16	+117	**		
2.83	Hay	5.11	+64	**		
1.30	Hay	4.07	+213	**		
4.30	Hay	4.33	0	NS	S nutritional effects eliminated. In year 1, gypsum caused a depressive effect due to induced Mg deficiency as a result of leaching	Sumner (unpublished data)
4.30	Hay	3.68	−14	NS		
4.30	Hay	3.86	−10	NS		
4.01	Hay	4.79	+19	*	S nutritional effects eliminated. Mg topdressing applied to all plots between first and second year	Sumner (unpublished data)
4.01	Hay	4.50	+12	NS		
4.01	Hay	4.99	+24	*		
19.37	Wet	27.56	+42	**		Riley and Brown (unpublished data)
10.54	Forage	10.58	+0	NS		
0	Hay	0			In year 1, trees were too young to yield. In years 2 and 3 late frost destroyed blooms	Sumner and Couvillon (unpublished data)
6.66 cm	Trunk diameter	7.30 cm	+10	NS		
7.27 cm	Trunk	7.69 cm	+6	NS		
6.29	Fruit	7.94	+26	*		
1.17	Hay	1.77	+51	**	First cut only	Sumner (unpublished data)
1.17	Hay	1.79	+53	**		
1.17	Hay	1.88	+61	**		
1.17	Hay	1.97	+68	**		
0.707	Lint	0.732	+4	NS		McCray and Sumner (unpublished data)
0.336	Lint	0.391	+16	*		
0.571	Lint	0.631	+11	**		

Table 8. *(Continued)*

Origin	Crop	Soil texture	Soil order	Gypsum rate (Mg ha^{-1})	Gypsum source	Year
11. Brazil	Apples		Inceptisol	4	Phosphogypsum	1
				4		2
			Oxisol	2		1
				2		2
12. Brazil	Beans		Oxisol	2.5	Phosphogypsum	1
				2.5		1
13. Brazil	Coffee		Oxisol	2.6	Phosphogypsum	1+2
				2.6		1+2
				0.5		1
				1–3		1+2
14. Brazil	Cotton		Oxisol	1.85	Phosphogypsum	1
				3.64		1
15. Brazil	Rice		Oxisol	1.4	Phosphogypsum	1
				1.4		1
				1.4		1
16. Brazil	Wheat		Oxisol	5.4	Phosphogypsum	1
17. Brazil	Corn		Oxisol	2.0	Phosphogypsum	1
18. Brazil	Corn		Oxisol	6.0	Phosphogypsum	1
				6.0		1
				6.0		1
				0		2
				0		2
				0		2

Table 8. *(Continued)*

Yield						
− Gypsum (Mg ha⁻¹		+ Gypsum	Change (%)		Remarks	Source
11.33	Fruit	11.66	+3	NS		Pavan et al. (1988)
7.33	Fruit	15.16	+107	**		
7.66	Fruit	8.33	+9	NS		
7.00	Fruit	11.33	+62	**		
0.72	Grain	1.10	+53		Wet season	Oliviera et al. (1986)
0.42	Grain	1.03	+145		Dry season	
1.02	Seed	2.00	+96		Average for 1980–1981	Guimraes (1986)
1.56	Seed	2.52	+61		Average for 1983–1984	
0.50	Seed	1.10	+120		Gypsum applied in hole	Freire et al. (1984)
1.00	Seed	2.21	+121		Average for 2 years	Guimraes et al. (1983)
2.61	Seed	2.34	−10			Rosolem and Machado (1984)
3.00	Seed	3.43	+14			Rosolem and Machado (1984)
1.5	Grain	1.6	+7			Oliviera et al. (1986)
1.1	Grain	1.1	0			
1.2	Grain	1.7	+42			
0.98	Grain	1.6	+63			Guimares (1986)
6.45	Grain	6.59	+2			Malavolta et al. (1986)
3.50	Grain	4.17	+19		Dry season with irrigation	Carvalho et al. (1986)
1.71	Grain	3.11	+82		Dry season with 21 days early drought	
0.27	Grain	0.72	+166		Dry season with 22 days late drought	
4.20	Grain	6.05	+44		Dry season with irrigation	
2.83	Grain	4.94	+74		Dry season with 25 days early drought	
2.67	Grain	4.63	+73		Dry season with 25 days late drought	

Table 8. *(Continued)*

Origin	Crop	Soil texture	Soil order	Gypsum rate (Mg ha^{-1})	Gypsum source	Year
19. Brazil	Beans		Ultisol	0.25	Phosphogypsum	1
			Oxisol	0.25		1
20. Brazil	Soybean		Ultisol	0.1	Phosphogypsum	1
			Ultisol	0.1		1
			Oxisol	0.1		1
			Oxisol	0.1		1
21. Brazil	Soybean		Oxisol	4.0		1
22. Brazil	*Brachiaria decumbens*		Oxisol	0.5	Phosphogypsum	1

most of the Georgian data, there has consistently been no response to gypsum in the first season, a fact that has important economic implications. In the few instances in which the effects of gypsum have been measured at different application rates (experiments 2, 3, and 6; Table 8), it is also clear that high rates of application were required. In these respects, there is a marked difference between these findings and those from Brazil, where in the first year responses are usually obtained to levels of application on the order of 1–2 Mg ha^{-1}. In all probability, these differences in behavior stem from the differences in precipitation (South Africa, 750–1,000 mm; Georgia, 1,000–1,500 mm; Brazil, 1,500–2,000+ mm) and soil properties.

In addition, in some of the Brazilian experiments (19, 20, and 22; Table 8), the responses to phosphogypsum are likely to arise from S and Ca nutrition, as the rates used were too low to appreciably affect the subsoils. The importance of rainfall is shown by the first-year response of alfalfa in experiments 7 and 9 (Table 8) from Georgia during a wetter-than-normal year. Rainfall differences do not provide a satisfactory explanation in all situations, however, and soil physical properties probably also play an important role. The South African soils studied, for example, are hydromorphic, and throughflow of water is certainly considerably slower than in the extremely well drained Brazilian soils. Soil data from experiment 2 (Table 8) indicate that vertical movement of gypsum at an application rate of 2 Mg ha^{-1} has been negligible even after 1,500 mm of rainfall over a 2-year period (Section III.C.1; Table 17). In addition, the possibility exists that Ca nutrition may play a more important role in Brazilian soils than is generally the case in South Africa and Georgia. Both topsoils and subsoils in the latter two localities have considerably higher Ca levels than are typically found in Brazil

Table 8. *(Continued)*

Yield – Gypsum (Mg ha^{-1})		Yield + Gypsum (Mg ha^{-1})	Change (%)	Remarks	Source
1.96	Grain	2.22	+13		Vitti et al. (1986)
1.10	Grain	1.70	+54		
1.31	Grain	1.79	+37		
1.26	Grain	1.61	+28		
1.13	Grain	1.62	+43		
1.26	Grain	1.61	+28		
2.82	Grain	2.89	+2	Wet season	Carvalho et al. (1986)
1.85	Hay	2.78	+50		
0.40	Daily gain	0.62	+55	Animal daily gain in kg/day	Vilela (1986)
69.1	Meat	161.4	+133	Total meat produced in kg/ha/yr	

Note. Level of statistical significance indicated by * (LSD > 0.05) and ** (LSD > 0.01). NS, not significant.

(Bouldin, 1979; and Table 7). Ritchey et al. (1982) have demonstrated the critical role of Ca in soils of the Brazilian Cerrados, and it seems possible that, in at least some instances, initial response to gypsum has resulted from Ca nutritional effects. Where the Ca supply in surface and upper subsoil horizons is marginal, a faster response to gypsum would be anticipated.

In general, relative response to gypsum has been greatest in seasons of moisture stress (indicated by low yields on plots not receiving gypsum), but it is significant that actual yield gains have been as large, even in seasons in which moisture stress was not particularly severe. As far as different types of gypsum are concerned, there appears to be relatively little to choose, with no meaningful differences being recorded other than to the presence of P in phosphogypsum (Section III.B.2). Fineness would, however, influence the solubility and reaction rate of gypsum (Section I.C). Fortuitously, the mined gypsum used in these studied has been very similar to phosphogypsum in terms of fineness.

2. Plant Composition

The nutrient content of plants is markedly affected by gypsum application. In all cases (Tables 9 and 10), the leaf Ca and S contents are elevated, always in the second and sometimes in the first year, after application, while the Mg content is significantly decreased (Carvalho et al., 1986; Farina and Channon, 1988b; Guimaraes, 1986; Pavan and Bingham, 1986; Quaggio et al., 1982; Sumner and Carter, 1988). Leaf K is often depressed by gypsum application. Such effects would not be unexpected, because gypsum applications markedly increase Ca and S availability and may exert an antagonistic effect on Mg and K absorption.

Table 9. Effect of gypsum on leaf nutrient composition of various crops

		Leaf nutrient content									
		N	P	K	Ca	Mg	S	Mn	Fe	Zn	
Crop	Treatment	(%)						(ppm)			Source
Apple	Control	2.62	0.21	1.50	1.02	0.33	0.22	150	100	26	Pavan et al. (1988)
	gypsum (4 Mg ha^{-1})	3.29	0.20	1.10	1.58	0.20	0.38	180	120	36	
Coffee	Control	3.60	0.14	3.30	0.38	0.18	–	400	–	36	Chaves, Pavan, and Miyazawa
	gypsum (3 Mg ha^{-1})	3.60	0.13	2.80	1.05	0.20	–	434	–	57	(unpublished data)
Corn	Control	0.83	0.10	0.82	0.12	0.17	0.04	108	–	–	Malavolta et al. (1986)
	gypsum	0.86	0.11	0.62	0.16	0.18	0.12	111	–	–	
Cotton	Control	–	–	1.59	1.57	0.32	–	–	–	–	Rosolem and Machado (1984)
	gypsum (5 Mg ha^{-1})	–	–	1.20	2.98	0.30	–	–	–	–	
Cotton	Control	–	–	1.55	3.59	0.58	–	–	–	–	Rosolem and Machado (1984)
	gypsum (5 Mg ha^{-1})	–	–	1.15	3.68	0.57	–	–	–	–	
Peanuts	Control	2.66	0.21	2.11	1.19	0.35	0.24	152	78	44	Quaggio et al. (1982)
	gypsum (2 Mg ha^{-1})	3.05	0.20	2.01	1.55	0.25	0.26	167	93	45	
Forage	Control	–	0.19	–	0.32	–	–	–	–	–	Vilela (1986)
	gypsum (0.5 Mg ha^{-1})	–	0.26	–	0.43	–	–	–	–	–	

Table 10. Effect of gypsum on nutrient content, DRIS indices, and yield of corn and alfalfa

Treatment		N	P	K	Ca	Mg	S	Mn	Zn	Cu	Yield	Source
		(%)						(mg kg^{-1})			(kg ha^{-1})	
Control	Value	2.80	0.27	2.31	0.32	0.39	–	80	26	11	4773	Farina and Channon (1988b)
	DRIS index	−2	−1	2	−20	16	–	6	0	1		1983/1984 season
Gypsum	Value	2.78	0.24	2.00*	0.38*	0.31*	–	93*	28	11	5811*	(corn)
	DRIS index	−2	−2	−1	−12	7	–	9	0	1		
Control	Value	2.74	0.25	1.88	0.22	0.24	–	77	27	12	7804	Farina and Channon (1988b)
	DRIS index	4	4	2	−34	7	–	9	4	5		1984/1985 season
Gypsum	Value	2.84	0.29	2.06	0.29*	0.24	–	82	26	12	9046*	(corn)
	DRIS index	0	2	0	−19	4	–	8	1	3		
Control	Value	2.56	0.25	2.40	0.32	0.40	0.18	67	17	12	5898	Farina and Channon (1988b)
	DRIS index	−2	0	6	−26	14	−2	6	−5	4		1985/1986 season
Gypsum	Value	2.60	0.26	2.35	0.45*	0.33*	0.22*	71	19	11	7046*	(corn)
	DRIS index	−4	−2	4	−9	7	−1	5	−5	1		
Control	Value	2.76	0.31	2.18	0.28	0.33	–	106	23	12	2668	Farina (unpublished data)
	DRIS index	−6	2	−1	−24	13	–	15	−7	3	3816*	1986/1987 season
Gypsum	Value	2.72	0.30	2.07	0.38	0.27	–	92	24	12		(corn)
	DRIS index	−6	1	−3	−9	5	–	10	−6	3		

(Continued)

Table 10. *(Continued)*

Treatment		N	P	K	Ca	Mg	S	Mn	Zn	Cu	Yield	Source
		(%)						(mg kg^{-1})			(kg ha^{-1})	
Control	Value	3.74	0.29	2.09	1.18	0.19	–	–	–	–	955	Sumner and Carter (1988)
	DRIS index	9	3	−6	−8	1	–	–	–	–		1985 Cut 1
Gypsum	Value	3.54	0.25	1.72	1.46*	0.18	–	–	–	–	2370*	(alfalfa)
	DRIS index	7	−4	−17	13	1	–	–	–	–		
Control	Value	3.97	0.33	2.50	1.14	0.25	–	–	–	–	731	Sumner and Carter (1988)
	DRIS index	7	3	−1	−20	11	–	–	–	–		1985 Cut 2
Gypsum	Value	4.00	0.31	2.32	1.47*	0.28	–	–	–	–	1709*	(alfalfa)
	DRIS index	1	−4	−9	−3	15	–	–	–	–		
Control	Value	3.74	0.37	2.47	0.91	0.25	–	–	–	–	789	Sumner and Carter (1988)
	DRIS index	8	14	1	−39	15	–	–	–	–		1985 Cut 4
Gypsum	Value	3.39*	0.30*	2.23	1.00*	0.23	–	–	–	–	1416*	(alfalfa)
	DRIS index	4	5	−1	−21	14	–	–	–	–		

Note. In experiment of Farina and Channon (1988a,b), all treatments received 15 Mg lime ha^{-1} mixed into the topsoil. In the experiment of Sumner and Carter (1988), all treatments received 3 Mg lime ha^{-1} in the topsoil. Both experiments received gypsum at the rate of 10 Mg ha^{-1}.

*Significant at 5% level.

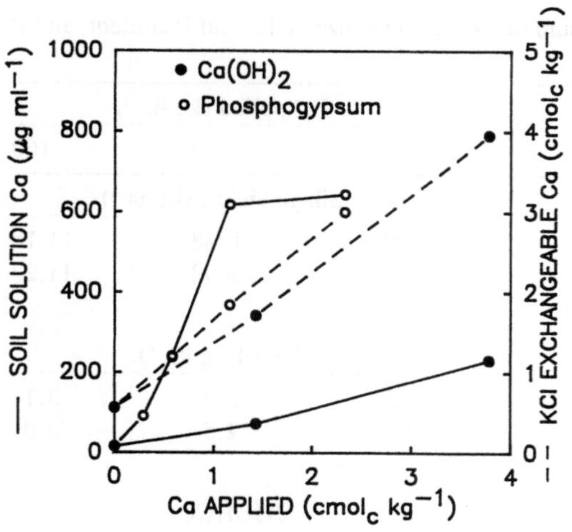

Figure 28. Effect of lime and phosphogypsum on soil solution and exchangeable Ca in an Appling soil (Sumner, unpublished data).

There is also general but less complete agreement that Mn uptake is increased following gypsum application. Increases in Ca uptake would be expected to reduce Mn absorption (Foy, 1984), and reasons for enhanced Mn uptake are not readily apparent. There is no evidence that the levels of extractable soil Mn are being increased by gypsum but, as discussed later in this section, soil solution Mn consistently increases after gypsum treatment. Perhaps more important, there is, as yet, no evidence that Mn uptake is elevated to near critical levels.

In most gypsum trials, adequate S nutrition has been ensured with small overall dressings of gypsum, and it has been possible to discount the effects on yield of enhanced S uptake. However, this has not always been true in the case of Ca nutrition, and there is evidence to suggest that part of the yield response to gypsum, even where subsoil Ca levels are considered to be adequate and topsoils have been heavily limed, is from improved Ca nutrition of the crop. This effect is illustrated in Table 10, in which both nutrient contents and DRIS (Diagnosis and Recommendation Integrated System) (Walworth and Sumner, 1987) indices have been presented. (For those readers not familiar with the DRIS approach to diagnosis, DRIS indices measure the relative balance between nutrients in tissue, with negative indices indicating relative insufficiency and positive indices relative excess.) In all cases of the nutrients considered, Ca was the most limiting by insufficiency, and when gypsum was applied, the requirement for Ca was lessened or eliminated with a concomitant significant yield increase, suggesting that a correct diagnosis was made in each case. Gypsum elevates the level of Ca in the soil solution more efficiently than lime (Figure 28), and this may contribute to root growth. It is also probable that increased Ca uptake would result from decreased Al availability (Huett and Menery, 1980), and decreased negativity of the Ca DRIS indices may be a reflection of decreased Al toxicity.

Table 11. Effect of P in phosphogypsum (0.74% P) on maize yield, leaf P content, and P soil test[a]

Phosphogypsum applied ($Mg\ ha^{-1}$)	P Applied ($kg\ ha^{-1}$)		
	40	80	160
	Silage yield ($Mg\ ha^{-1}$)		
5	10.07	10.68	11.18
0	9.16	9.92	11.22
LSD (0.05) at a given P rate = 0.51			
	Leaf P ($g\ kg^{-1}$)		
5	2.5	2.9	3.1
0	2.2	2.5	3.0
LSD (0.05) at a given P rate = 0.1			
	$NH_4HCO_3^-$ Extractable soil P ($mg\ l^{-1}$)		
5	13	24	45
0	8	18	44
LSD (0.05) at a given P rate = 4			

From Farina (unpublished data).

[a]0- to 0.15-m soil samples collected 3 weeks after gypsum application.

In cases where PG is used on P-deficient soils, its P content, which may range from 0.2 to 1.0%, can be of great benefit to crop growth. This effect, which has been quantitatively measured, is illustrated in Table 11. First-year yield response and changes in leaf and soil P as affected by PG were only obtained at the lowest two rates of P applied. This experiment also included three lime rates in factorial combination with P, and nonsignifiance of the lime × gypsum interaction (the effects were clearly additive) suggested that the gypsum effects measured were not related to Al. Thus, the P content of PG confers added value to this product, which warrants consideration when evaluating the economics of its use as a subsoil ameliorant. In the case illustrated (Table 11), 5 Mg ha^{-1} of PG was approximately equivalent to 40 kg P ha^{-1} in terms of P nutrition.

Souza and Ritchey (1986) found that improved root development following gypsum use also resulted in enhanced nitrate recovery from subsoil horizons. Total N uptake by maize increased from 91 to 135 kg ha^{-1} as a result of a 6 Mg ha^{-1} application of gypsum. It is probable that the high N applications generally used in other gypsum studies have precluded measurement of a similar effect. In addition, Brazilian soils tend to be extremely well drained and to have a lower topsoil organic matter content than many acid soils in South Africa and the southeastern United States. This is an area deserving closer examination, as improved N use efficiency would significantly improve the overall economics of subsoil amelioration and retard re-acidification.

Table 12. Effect of gypsum on alfalfa root distribution in an Appling soil profile

Depth (m)	Control Root mass (g m^{-3})	Control Root length (m m^{-3})	Gypsum Root mass (g m^{-3})	Gypsum Root length (m m^{-3})
0–0.15	124.2	375.3	115.0	439.1
0.15–0.30	14.1	39.7	30.0	94.3
0.30–0.45	6.1	11.2	18.5	95.5
0.45–0.60	8.8	52.0	10.1	112.5
0.60–0.75	1.5	4.3	5.7	28.3
0.75–0.90	0.0	2.4	8.5	35.6
0.90–1.05	0.0	0.6	0.3	0.8
Total (0–1.05 m)	154.7	484.7	188.1	806.2
Total (0.15–1.05 m)	30.5	109.4	73.1	367.1

Reprinted from Sumner and Carter, 1988, by courtesy of Marcel Dekker, Inc.

3. Root Responses

Relatively few root studies have been performed on the experiments presented in Table 8, largely because of the very laborious nature of such work. However, sufficient detailed studies have been carried out to clearly document the beneficial effects of gypsum on subsoil root penetration. Sumner and Carter (1988) studied the effect of gypsum on alfalfa root distribution in the profile of a Georgia Ultisol (Table 12). Gypsum treatment resulted in substantial penetration of roots to deeper than 1 m, whereas in the control plots few roots penetrated beyond 0.6 m. On both a mass and length basis, the quantity of roots in the subsoil of the gypsum treatment exceeded that of the control three- to fivefold, which is reflected in the alfalfa response reported in Table 8 for experiment 5. Reports corroborating this type of response have been published for apples (Pavan and Bingham, 1986), maize (Carvalho et al., 1986; Farina and Channon, 1988b; Souza and Ritchey, 1986), and beans and rice (Oliveira et al., 1986). In all cases, the gypsum treatment increased the Ca level in the subsoil, decreased the Al level, and promoted root growth and yield.

In a number of cases, a relative decrease in root growth has been observed at intermediate depths in the subsoil (Farina and Channon, 1988b; Souza and Ritchey, 1986; Sumner and Carter, 1988). This effect is illustrated in Tables 12 and 13. In the case of the alfalfa experiment (Table 12), decreased root growth occurred where the bulk density of the soil reached its highest value (Sumner et al., 1986), and it is not possible to separate physical from chemical effects. In the case of the maize trials (Table 13), however, there was no evidence of physical

Table 13. Effect of gypsum application on root distribution of maize and apple in the profiles of highly weathered soils

Depth	Maize[a] Root density ($m\ l^{-1}$)		Maize[b] Relative root distribution (%)		Apple[c] Root density ($cm\ g^{-1}$)	
(m)	Control	Gypsum	Control	Gypsum	Control	Gypsum
0–0.15	3.10	2.95	53	34	50	119
0.15–0.30	2.85	1.60	27	25	60	104
0.30–0.45	1.80	2.00	10	12	18	89
0.45–0.60	0.45	3.95	8	19	18	89
0.60–0.75	0.08	2.05	2	10	18	89

[a] From Farina and Channon (1988b).
[b] From Souza and Ritchey (1986).
[c] From Pavan (unpublished data).

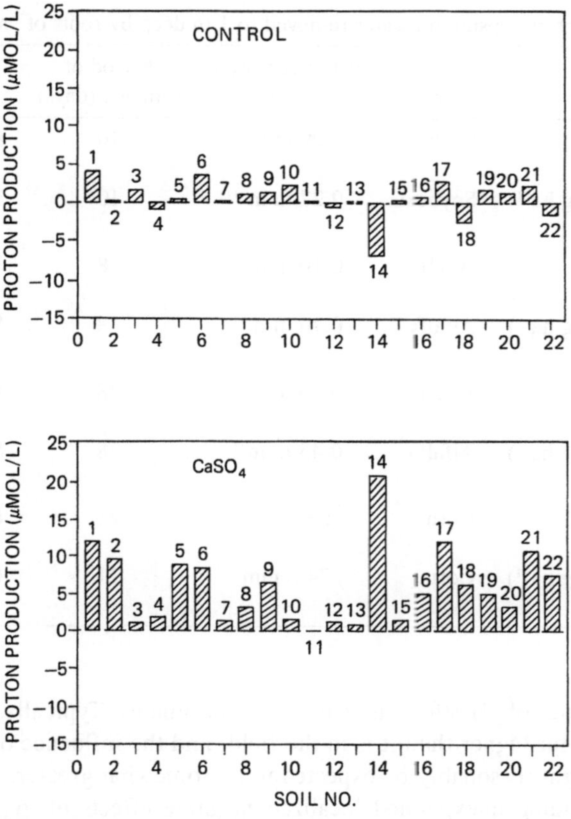

Figure 29. Rhizosphere acidification in alfalfa seedling grown in pots with (A) and without (B) the addition of 1 g kg^{-1} $CaSO_4{\cdot}2H_2O$. Proton production calculated as an index from the pH difference between bulk soil and soil removed from root surfaces, measured in H_2O at a soil/solution ratio of 1:2 (Fey, Noble, and Sumner, unpublished data).

impedance to root growth, and the depression was clearly related to gypsum application. At present there is no entirely satisfactory explanation for this effect or, for that matter, any measure of its influence on crop performance.

It is, however, not unreasonable to assume that such an effect would lessen the overall benefits potentially obtainable from gypsum, which is an area deserving further study. Because the zone of relative root growth depression coincides with the zone of maximum SO_4^{2-} accumulation, anion antagonism has been invoked as a possible causal mechanism. Plant analytical data, however, have not supported this suggestion. A more plausible mechanism may involve gypsum-induced depressions in rhizosphere pH (Figure 29). Presumably, such depressions resulted from enhanced cation absorption, and it is conceivable that toxic levels of Al can exist at the root surface even though the levels of Al in the bulk soil are acceptable. It is possible that this effect has also been partly responsible for the

Table 14. Effect of gypsum on water removed to 1 m deep by roots of various crops

Treatment	Crop	Water removed to 1 m	Period of removal (days)	Source
Control	Soybean	0.09 m m^{-1}	10	Hammel et al. (1985)
Gypsum (35 Mg ha^{-1})	Soybean	0.27 m m^{-1}	10	Hammel et al. (1985)
Control	Alfalfa	0.10 m m^{-1}	8	Sumner et al. (1986)
Gypsum (10 Mg ha^{-1})	Alfalfa	0.16 m m^{-1}	8	Sumner et al. (1986)
Control	Alfalfa	0.36 m m^{-1}	28	Sumner et al. (1986)
Gypsum (10 Mg ha^{-1})	Alfalfa	0.43 m m^{-1}	28	Sumner et al. (1986)
Control	Corn	53.5 mm	21	Carvalho et al. (1986)
Gypsum (6 Mg ha^{-1})	Corn	74.8 mm	21	Carvalho et al. (1986)

apparent absence of pH effects in many pot experiments. Typically, rooting density in pots is far higher than it is in the field, and the influence of rhizosphere properties might reasonably be expected to be somewhat greater.

Notwithstanding unexplained localized negative effects of gypsum on root proliferation, the overall improvement in subsoil root development has obvious benefits in overcoming the deleterious effects of seasonal droughts.

4. Water Usage

Confirmatory evidence for the benefits of deeper rooting following gypsum use is to be found in water extraction patterns down the profile. Such data (Table 14) indicate that when the gypsum front has penetrated the subsoil, roots are able to

Table 15. Effect of gypsum on leaf water potential of corn

Treatment	Crop	Leaf water potential bars	Source
Control	Corn	−7.76	Sumner (unpublished data)
Gypsum (35 Mg ha^{-1})	Corn	−4.59	Sumner (unpublished data)
Control	Corn	−17	Carvalho et al. (1986)
Gypsum (10 Mg ha^{-1})	Corn	−13	Carvalho et al. (1986)
Control	Corn	−26	Carvalho et al. (1986)
Gypsum (10 Mg ha^{-1})	Corn	−22	Carvalho et al. (1986)

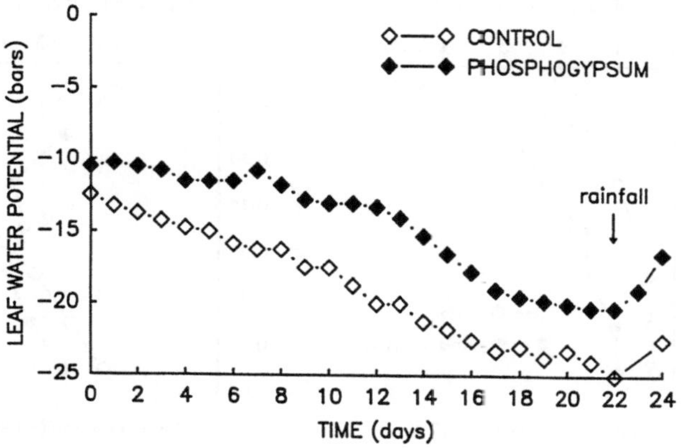

Figure 30. Leaf water potentials for apples grown under moisture stress on untreated (control) and phosphogypsum-treated Inceptisols (Pavan, unpublished data).

extract more water than in control treatments. This increased water uptake allows the plant to better overcome drought stress and therefore accumulate dry matter more successfully on the gypsum-treated soils. This improved water uptake is corroborated by leaf water potential measurements (Table 15 and Figure 30).

C. Effects on Soil Properties

1. Exchangeable Plant Nutrients and Acidity

a. Basic Cations

Treatment of soil with gypsum has marked effects on the chemical properties of acid, highly weathered profiles. This is illustrated in Figures 31, 32, and 33 for soils from Georgia, South Africa, and Brazil, respectively. As expected, exchangeable Ca is increased uniformly down the profile after sufficient time has elapsed for the surface-applied gypsum to dissolve and move in the percolating rainwater. Magnesium is reduced in the upper part of the profile by gypsum application and accumulates in the lower portion, which presumably is simply an intermediate stage in its removal from the profile. This is corroborated by data from a sandy soil in which gypsum resulted in removal of Mg to very low levels to a depth of 1 m (Syed, 1987). On the other hand, the behavior of K as a result of gypsum application appears to differ according to location. In Georgia (Figure 31), K in the upper part of the profile is markedly reduced and does not appear to accumulate in exchangeable form in the subsoil. As yields on the gypsum treatment were consistently higher than the control, differential uptake of K could account for some of the observed difference, or some fixation or immobilization of K may be taking place. In the South African soil, little change in exchangeable K occurs down the profile (Figure 32), whereas in Brazil (Figure 33) the K behaves in a manner similar to that in the Georgia soil. Other studies by Souza

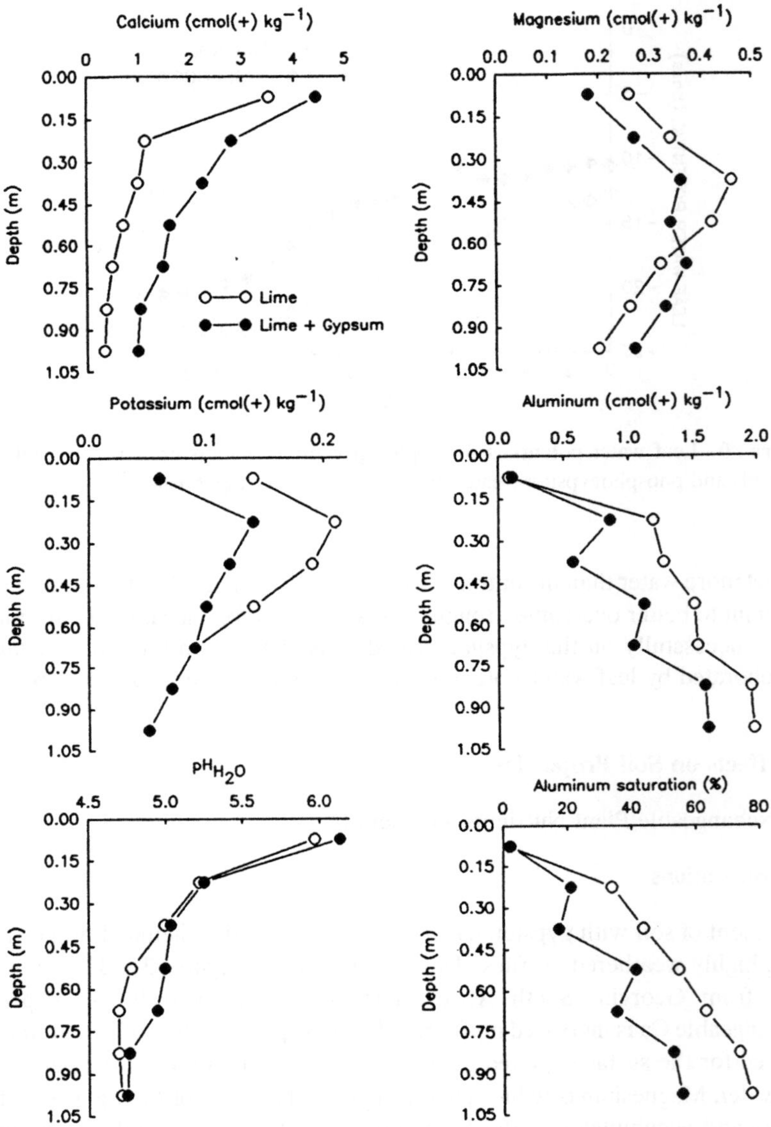

Figure 31. Effect of gypsum on soil chemical properties for an Appling (Georgia) soil. Lime and 10 Mg ha^{-1} PG incorporated to 0–0.15 m depth 5 years previously (Sumner, unpublished data).

and Ritchey (1986) on another Brazilian soil suggest that K behaves in a similar manner to Mg. Independent studies on the competitive exchange of Ca with K and Mg (M.A. Pavan, unpublished data) indicate that a Brazilian soil had a higher selectivity coefficient for K than Mg when in competition with Ca, which is similar to its behavior in the field. The behavior of K in the South African soil may also be caused by a higher affinity for K than Mg.

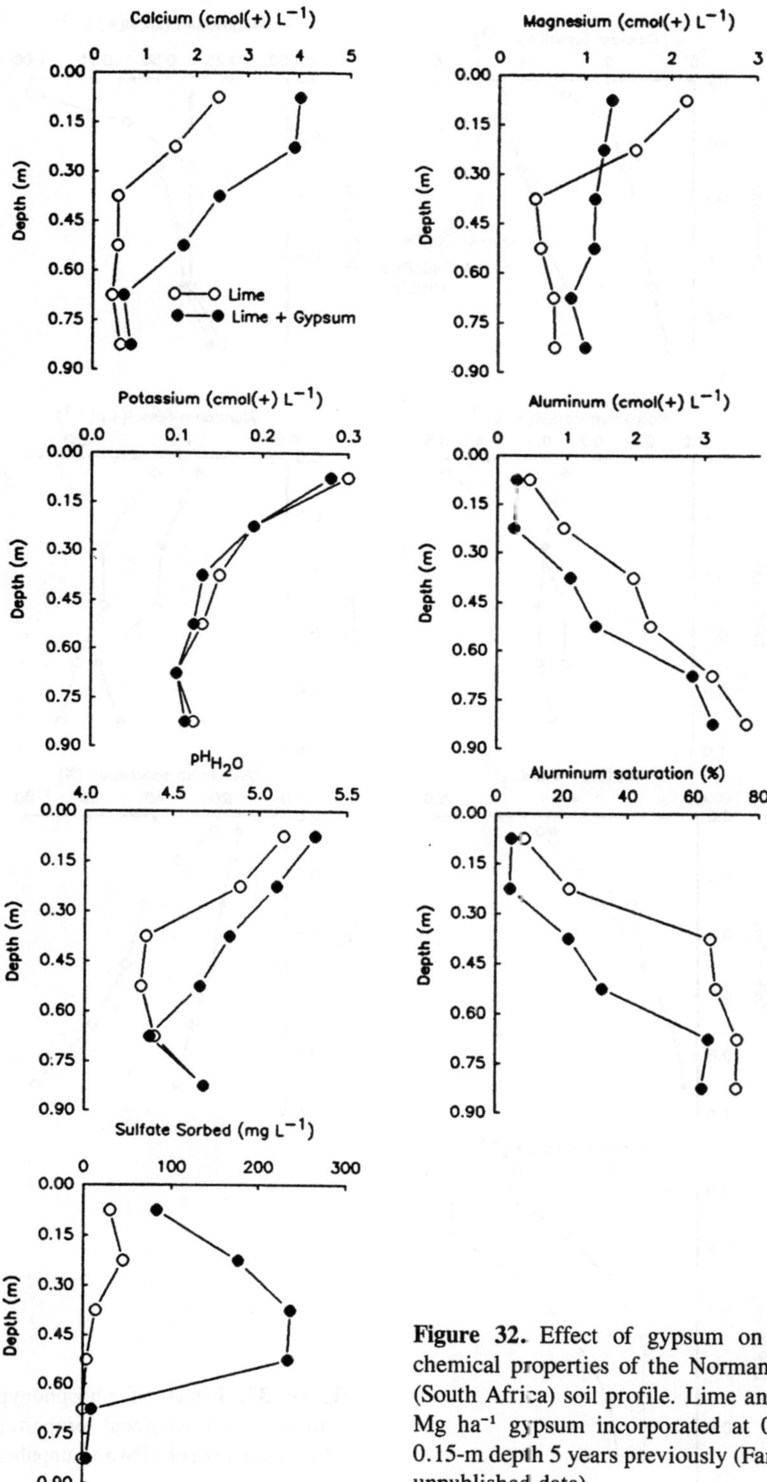

Figure 32. Effect of gypsum on soil chemical properties of the Normandien (South Africa) soil profile. Lime and 10 Mg ha^{-1} gypsum incorporated at 0- to 0.15-m depth 5 years previously (Farina, unpublished data).

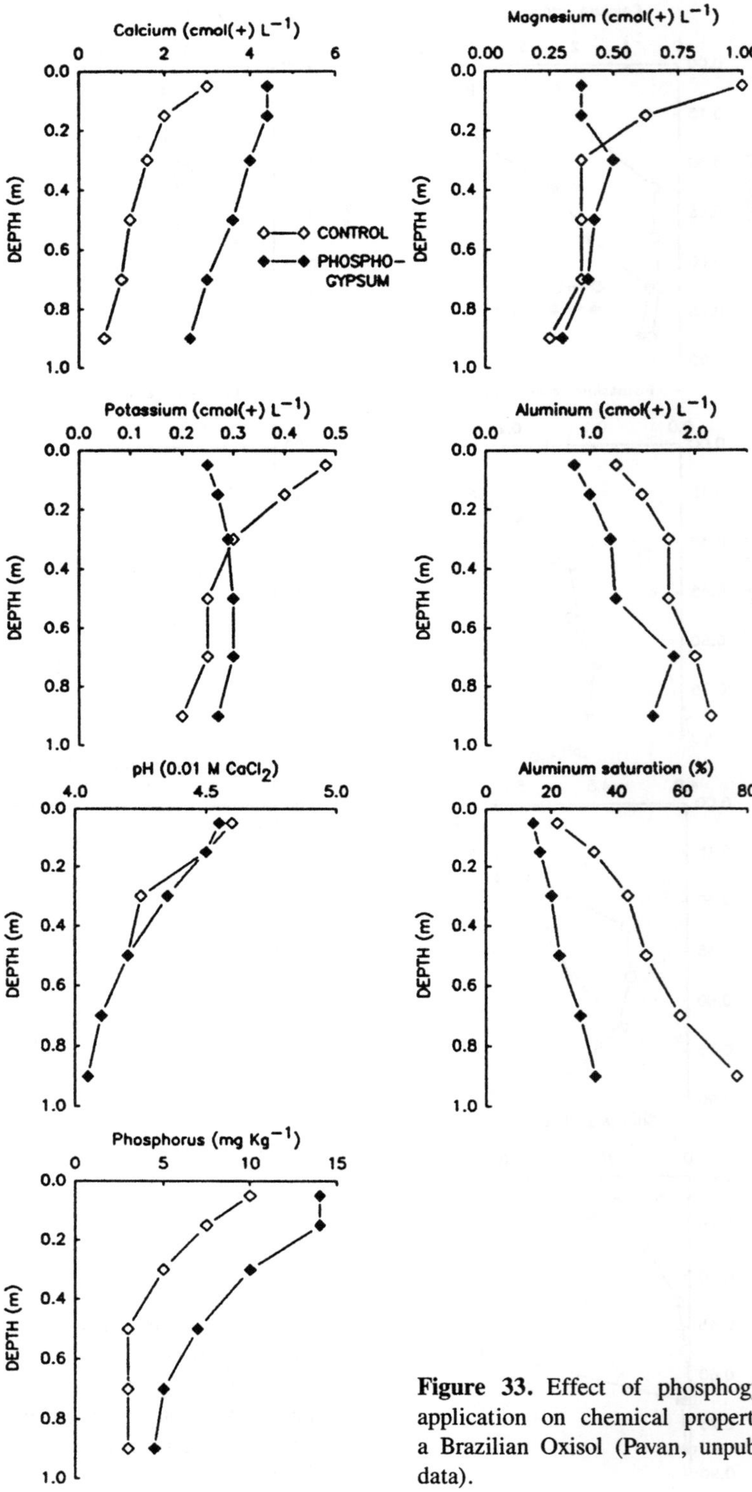

Figure 33. Effect of phosphogypsum application on chemical properties of a Brazilian Oxisol (Pavan, unpublished data).

b. Acidity and Aluminum

In the case of Al, the pattern is essentially the same at all three locations, with the gypsum treatment reducing both the exchangeable Al and degree of Al saturation down the entire profile, with no evidence of Al moving ahead of the advancing Ca front as was found for Mg (Figures 31, 32, and 33). Leaching studies (O'Brien and Sumner, 1988; Sumner et al., 1985; Buyeye et al., 1985) substantiate this observation, and indicate that little if any measurable Al is removed in the leachate from reconstituted soil profiles. This contrasts with the finding of Oates and Caldwell (1985), who used very short subsoil columns with complete mixing of gypsum. Sumner (1970) and Kotze and Deist (1975) also reported the leaching of some Al from soil columns treated with gypsum. Under field and laboratory conditions, gypsum application has been observed to increase (Figures 31, 32, and 33; and Couto et al., 1979; Farina and Channon, 1988b; Hue et al., 1985; Ritchey et al., 1980; Souza and Ritchey, 1986), decrease (Pavan et al., 1982, 1984; Quaggio et al., 1982; Rosolem and Machado, 1984), or have little effect on (Pavan et al., 1982, 1984) soil pH down the profile.

The magnitude of the pH change, where it is observed, it usually small (of the order of 0.2–0.3 pH units), and typically is not detectable in an electrolyte suspension but only when pH is measured in water. This contradictory picture may arise as a result of the envisaged conflicting reactions between gypsum and soil surfaces, in which Ca replaces H and Al (which hydrolyzes to give H^+) and SO_4 replaces OH by ligand exchange. The resultant pH measured will thus depend on the extents of the two reactions in any particular case. Pavan et al. (1984) showed that, in soils high in exchangeable Al, H^+ release by hydrolysis was likely to exceed OH^- release, causing soil pH to decrease, whereas in soils that are highly weathered but low in exchangeable Al, the reverse was likely to occur (Ritchey et al., 1980). The results of Kiehl and Franco (1984) confirmed this behavior. When the effect of the first reaction is held constant by measuring pH in $CaCl_2$ and $CaSO_4$ solutions of equivalent Ca, the pH of the soil is raised by gypsum as a result of the greater replacing power of SO_4 than Cl for OH groups on the surface (Table 16). Thus, provided that the salt effect on pH measurement is relatively minor, gypsum will usually raise soil pH slightly. This will occur particularly in highly weathered soils which have larger quantities of hydroxylated surfaces on sesquioxides available for exchange, as well as a low CEC and therefore exchangeable acidity. The largest increases in pH (Figures 31 and 32) usually coincide with the zone of maximum sulfate sorption. Because of variability under field conditions, however, this effect cannot always be measured accurately.

c. Anionic Nutrients

As far as anions are concerned, sulfate from gypsum is relatively strongly retained in the subsoils of highly weathered soils (Figure 32), with NO_3^- being displaced and removed from the profile (Souza and Ritchey, 1986). Sulfate does not appear in the leachate from soil columns until after all the Cl^- has been removed (Sumner et al., 1985), and NO_3^- would presumably behave in a manner similar to

Table 16. Effect of nature and concentration of electrolyte on pH of Appling soil

Concentration or ionic strength (mol l^{-1})	pH at given concentration in			pH at given ionic strength		
	$CaSO_4$	$CaCl_2$	ΔpH	$CaSO_4$	$CaCl_2$	ΔpH
0.0500	–	–	–	4.35	3.93	0.45
0.0300	–	–	–	4.54	4.20	0.34
0.0140	4.42	4.14	0.28	4.72	4.44	0.28
0.0014	4.82	4.62	0.20	5.08	4.91	0.17
0.0007	4.93	4.82	0.11	5.19	5.06	0.13

Note: $\Delta pH = pH_{CaSO_4} - pH_{CaCl_2}$.
From Sumner et al. (1986).

Cl^-. Liming topsoil increases the movement of SO_4^{2-} into the subsoil (Table 17). Buyeye et al. (1985) have reported an identical effect in leached columns. Specific adsorption of much of the SO_4^{2-} leached into the subsoil is indicated by the commonly much lower quantity of SO_4^{2-} that can be extracted with $CaCl_2$ than with Ca phosphate solution (Farina and Channon, 1988b; Buyeye et al., 1985).

Although it might be anticipated that dissolved gypsum will enhance the solution activity of soil phosphate, several attempts to measure enhanced extractable P have indicated that there is no measurable effect of gypsum on P solubility (Buyeye and Fey, 1987). On the other hand, there is evidence from column leach-

Table 17. Effect of surface lime applications on movement of calcium phosphate-extractable sulfate

Gypsum applied (Mg ha^{-1})	Mg extractable SO_4^{-2} L^{-1}			
	0 mg lime ha^{-1} (pH 4.70)	3 mg lime ha^{-1} (pH 5.11)	6 mg lime ha^{-1} (pH 5.55)	12 mg lime ha^{-1} (pH 6.00)
	0.15- to 0.30-m horizon [LSD (0.05) = 73]			
0	33	34	50	54
1	42	58	79	59
2	52	75	91	84
4	110	128	120	160
8	213	225	253	205
16	241	260	247	587
	0.30- to 0.45-m horizon [LSD (0.05) = 35]			
0	8	4	3	7
1	7	5	5	7
2	5	11	6	7
4	15	6	15	11
8	33	37	34	37
16	28	65	68	123

From Farina (unpublished data).

ing studies to indicate that Si is often mobilized by gypsum application (Lemus-Grob, 1985; Sumner et al., 1985; O'Brien and Sumner, 1988).

d. Micronutrients

There has been very little investigation of the effect of large gypsum applications on micronutrients. Some studies have found evidence for a moderate displacement of Zn from the topsoil, but the effect was not consistent (M.V. Fey, unpublished data). The data in Tables 9 and 10 suggest that micronutrient cation availability to plants is not detrimentally affected by gypsum application.

2. Soil Solution Composition

The effect of gypsum application on the composition of the soil solution removed from a number of soils is shown specifically with respect to Ca, Al, and acidity levels in Table 18, and more generally in Table 19. Most of the data in Table 18 pertain to rewetted core samples from field plots treated with gypsum, whereas Table 19 refers to an incubation of soil at field capacity in closed pots. Viewed as a whole, Table 18 shows that after gypsum treatment: (1) sometimes, but not always, the solution pH increases; (2) the total Al concentration usually increases, most of the increase being in the form of the soluble $AlSO_4^+$ complex; (3) the concentration and activity of Al^{3+} invariably both decrease, the latter especially; the (4) the concentration and activity of Ca^{2+} increases markedly in all cases.

The broader picture of ionic interactions only emerges, however, from the data in Table 19. Here, pH does not increase, as it does in the field subsoils infiltrated by gypsum through leaching (Figures 31 and 32; Table 18); presumably in these closed pots with direct addition of gypsum and no leaching, the "salt effect" on pH predominates over the "sulfate effect". The two clay soils (Appling, Normandien) show a very subdued increase in Ca, SO_4, and total electrolyte concentration after gypsum addition, by comparison with the Tifton sand. This shows that large salt sorption has taken place, presumably associated with ionic strength-dependent variable charge on sesquioxide surfaces.

In the Tifton sand, which is electrolytically swamped by the 2 Mg ha^{-1} gypsum addition, exchange appears to have been the dominant process in giving rise to markedly greater concentrations of native ions in solution after gypsum addition. On the other hand, in the Appling and Normandien subsoils, which exhibit the greatest salt-sorbing tendencies, gypsum addition has actually reduced the concentration of both Al (despite a decrease in solution pH) and K, even though Mg, Na, and other ion concentrations have generally increased in keeping with the trend to be expected if ion exchange were to dominate the soil–gypsum reaction. Thus a strong sorption of both Ca and SO_4 and a cosorption of K and Al, without the release of a corresponding quantity of other ions into solution, has taken place with these subsoils. Some of the other effects of gypsum shown in Table 19 are worth noting: these include a tendency for soluble Mn, Si, No_3, Cl, and F levels to increase, and for soluble PO_4 to decrease slightly. These trends are more pronounced in the clay soils than in the Tifton sand.

Table 18. Effect of gypsum on the Ca, Al, and acidity status of the soil solution of highly weathered soils[a]

Treatment	Sample identification	Soil solution pH	Concentration (μM)								Activity (μM)		Source
			Al_t	Al^{3+}	$AlSO_4^+$	$AlOH^{2+}$	$Al(OH)_2^+$	$Al(OH)_3^0$	Ca_t	Ca^{2+}	Al^{3+}	Ca^{2+}	
Control	Ultisol (0.15–0.30 m)	4.78	13.0	8.4	0.5	3.2	0.9	0.0	420	410	4.2	344	Sumner et al. (1986)
	Ultisol 0.45–0.60 m)	4.80	11.0	7.8	0.0	2.4	0.5	0.3	250	250	4.0	114	
	Ultisol (0.74–0.90 m)	4.87	9.0	5.8	0.0	2.4	0.5	0.3	70	70	3.0	37	
Gypsum (10 Mg ha^{-1})	Ultisol (0.15–0.30 m)	4.85	15.0	5.4	5.0	2.9	1.0	0.7	4220	3790	2.4	2537	Sumner et al. (1986)
	Ultisol (0.45–0.60 m)	4.80	14.0	7.3	3.3	2.4	0.6	0.4	1620	1510	3.0	1123	
	Ultisol (0.75–0.90 m)	4.80	12.0	6.4	0.3	3.3	1.2	0.8	750	740	3.3	606	
Control	Oxisol (soil #2)	4.1	51.3	43.4	2.6	4.4	0.0	0.0	180	180	–	–	Pavan et al. (1982)
$CaSO_4$[b]	Oxisol (soil #2)	4.2	58.9	30.2	25.3	3.4	0.0	0.0	2550	1770	–	–	Pavan et al. (1982)
Control	Ultisol (soil #3)	4.4	58.9	31.9	2.5	6.0	0.9	0.0	250	250	–	–	Pavan et al. (1982)
$CaSO_4$[b]	Ultisol (soil #3)	4.5	70.8	26.0	25.6	5.5	1.4	0.0	2500	1750	–	–	Pavan et al. (1982)
Control	Oxisol (topsoil)	3.98	16.2	136.1	2.5	5.6	1.5	0.0	2870	2785	44.4	1678	Fey, Noble & Sumner (unpub)
$CaSO_4$ (2 Mg ha^{-1})[b]	Oxisol (topsoil)	4.11	26.9	128.7	116.0	7.0	2.7	0.0	9700	7772	28.6	3869	Fey, Noble & Sumner (unpub)

Control	Oxisol (subsoil)	4.60	8.91	3.40	0.01	1.0	1.6	0.1	140	140	2.60	107	Fey, Noble & Sumner (unpub)
$CaSO_4$ (2 Mg ha^{-1})[b]	Oxisol (subsoil)	4.71	6.0	1.50	1.2	0.3	0.6	0.0	4610	3818	0.40	2048	Fey, Noble & Sumner (unpub)
Control	Ultisol	4.1	34	23.1	8.5	2.4	–	–	690	–	14	470	Buyeye et al. (1985)
$CaSO_4$ (3 Mg ha^{-1})	(15–20-cm layer, from	4.5	38	20.9	12.2	4.9	–	–	2000	–	11	1200	Buyeye et al. (1985)
$CaSO_4$ (6 Mg ha^{-1})	leached column)	4.5	55	23.7	27.0	4.4	–	–	3600	–	12	1800	Buyeye et al. (1985)
$CaSO_4$ (9 Mg ha^{-1})		4.4	69	26.9	40.0	4.1	–	–	5100	–	13	2400	Buyeye et al. (1985)
$CaSO_4$ (12 Mg ha^{-1})		4.5	63	24.6	35.7	3.8	–	–	4700	–	11	2200	Buyeye et al. (1985)
Control	Oxisol (0.2–0.4 m)	–	23.0	16.1	0.5	4.8	1.2	0.5	40	40	–	–	Pavan et al. (1984)
	Oxisol (0.6–0.8 m)	–	50.0	40.0	0.5	8.0	1.5	0.0	60	60	–	–	Pavan et al. (1984)
PG	Oxisol (0.29–0.40 m)	–	53.0	25.4	21.2	5.3	1.1	0.0	1800	680	–	–	Pavan et al. (1984)
	Oxisol (0.6–0.8 m)	–	60.0	36.0	17.4	6.0	1.2	0.0	1000	720	–	–	Pavan et al. (1984)

[a] Ion speciation and activity calculation according to GEOCHEM (Sposito and Mattigod, 1979).

[b] Pure $CaSO_4 \cdot 2H_2O$.

Table 19. Composition of the soil solution after incubating soils in pots for 3 months at field capacity with or without 1000 mg kg^{-1} of $CaSO_4{\cdot}2H_2O$

Soil	Horizon	Treatment	pH	Cations (mmol l^{-1})							Anions (mmol l^{-1})						Electrolyte concentration
				Ca	Mg	K	Na	NH_4	Al	Mn	SO_4	NO_3	Cl	PO_4	F	Si	[mmol(c) l^{-1}]
Appling	Topsoil	Control	4.33	1.68	0.55	0.33	0.84	0.17	.008	0.015	0.13	4.53	0.42	0.005	0.020	0.259	5.6
	(loam)	Gypsum	4.35	8.47	1.06	0.36	1.11	0.14	.071	0.251	6.95	6.30	0.40	0.006	0.055	0.340	20.9
	Subsoil	Control	5.31	0.02	0.01	0.10	0.91	0.17	.005	ND	0.12	0.22	0.27	0.004	0.037	0.218	1.0
	(clay)	Gypsum	4.87	0.83	0.22	0.04	1.58	0.17	.003	0.002	1.12	0.71	0.42	0.002	0.041	0.392	3.9
Normandien	Topsoil	Control	4.38	1.20	1.29	0.98	1.75	0.68	.013	0.017	0.04	8.40	0.42	0.008	0.023	0.266	8.4
	(clay)	Gypsum	4.28	4.86	2.18	0.92	1.94	0.93	.021	0.056	3.97	10.10	0.50	0.004	0.058	0.301	17.9
	Subsoil	Control	5.05	0.03	0.02	0.21	1.20	0.13	.013	ND	0.02	0.84	0.41	0.005	0.035	0.146	1.4
	(clay)	Gypsum	4.63	0.67	0.65	0.17	3.08	0.15	.001	0.002	1.11	1.90	0.61	ND	0.056	0.247	4.8
Tifton	Topsoil	Control	6.00	3.00	0.99	1.33	0.68	0.17	.011	0.001	0.77	5.04	0.46	0.020	0.011	0.141	9.0
	(sand)	Gypsum	5.74	14.41	3.27	2.15	1.09	0.16	.015	0.001	12.85	8.00	0.54	0.014	0.078	0.140	38.8
	Subsoil	Control	4.72	0.15	0.04	0.46	0.70	0.19	.014	0.002	0.29	0.63	0.57	0.007	0.003	0.287	1.7
	(sand)	Gypsum	4.13	14.43	0.60	0.93	1.21	1.12	.700	0.061	13.29	0.40	0.40	0.001	0.085	0.333	35.2

Note. Inductively coupled plasma spectroscopy for P, Si, and cations other than NH_4 (by colorimetry); ion chromatography for remaining anions; solution removed by centrifuge filtration (<0.2 m). ND, not detected.

From Fey, Noble, and Sumner (unpublished data).

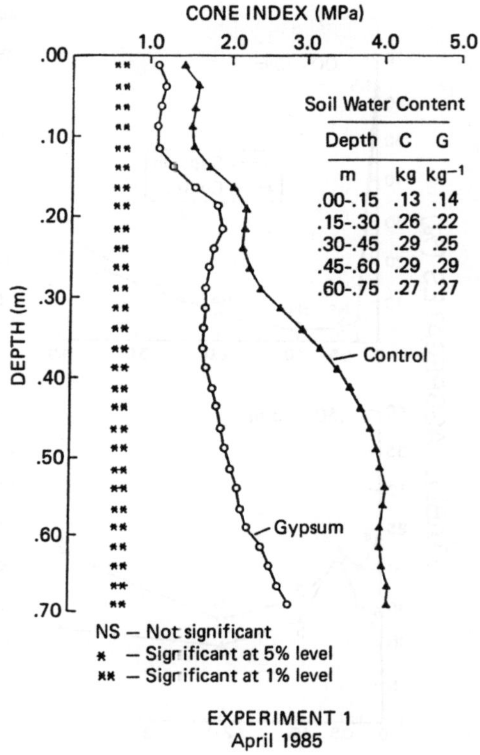

Figure 34. Effect of gypsum (10 Mg ha^{-1}) on penetration resistance (cone index) for Appling soil profiles (Radcliffe et al., 1986).

In Section III.F, an attempt is made to isolate from these soil responses those which are most likely to be connected with the ameliorative effect of gypsum on plant growth.

3. Physical Properties

The effect of surface-applied gypsum on the physical properties of Bt horizons in Georgia Ultisols was studied by Radcliffe et al. (1986) using mechanical impedance measured with a cone penetrometer as an indicator of the strength of hardpan layers. Four years after application of gypsum on the surface, there was a highly significant reduction in the cone index (resistance to penetration) (Figure 34) down the entire profile, but the difference was most marked below a depth of 0.30 m, where there was a highly significant increase in large water-stable aggregates (Figure 35). These improved physical conditions were attributed to an improvement in soil structure following greater subsoil root activity in response to the chemical amelioration produced by gypsum. There is also the possibility that Ca and ionic strength directly affect clay behavior, causing a change in sheer modulus, which might be expected to influence the penetrometer resistance.

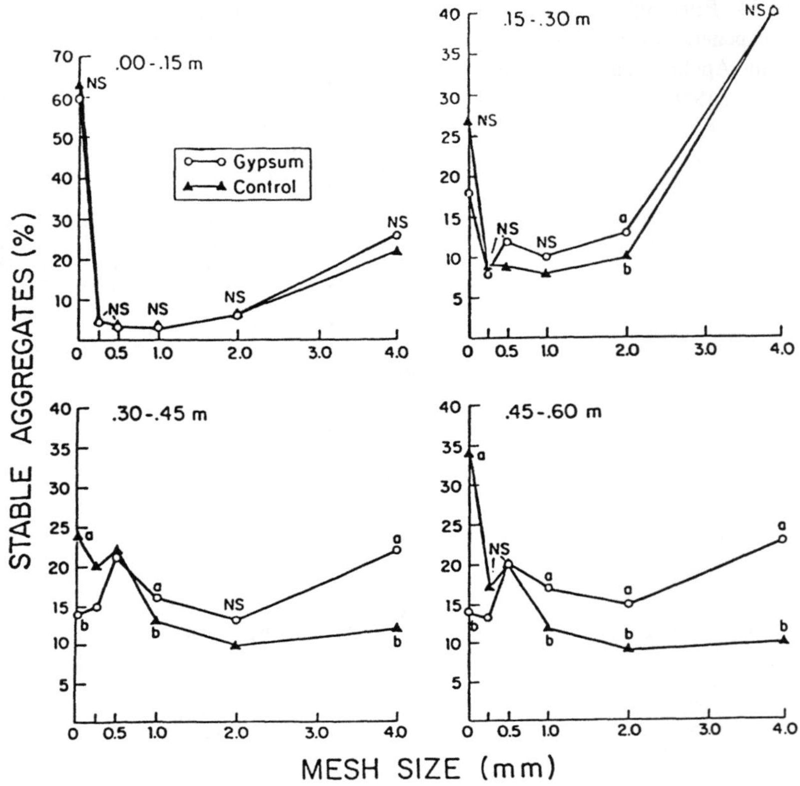

Figure 35. Effect of gypsum (10 Mg ha^{-1}) on water-stable aggregates in Appling soil profiles (Radcliffe et al., 1986).

An apparent negative effect of gypsum on the physical properties of an Al-rich acid Oxisol has been reported by Roth and Pavan (1988), who showed that water infiltration was decreased by liming and to a much lesser extent by gypsum application. This observation apparently contradicts the positive effects of gypsum on acid soils reported earlier (Section II.D). The soil used by Roth and Pavan (1988) had not been limed as had those referred to in Section II.D (Miller, 1987a). Thus the addition of PG may have resulted in a displacement of Al, causing a less flocculated condition that reduced infiltration. Had gypsum been added to the limed soil, a positive effect may possibly have been obtained, in accordance with the fact that liming had drastically reduced infiltration. An alternative explanation could be that specific adsorption of sulfate may impart sufficient negative charge to create increased double layer repulsion, resulting in a more dispersed condition. The long-term effect of this gypsum application was reported by Roth et al. (1984) who, 2 years after gypsum application, found higher aggregate stability and infiltrability in the gypsum treatment. The long-term positive effect was ascribed to improved biological activity as a result of improved Ca and P nutrition, which in time more than compensated for the initial negative effect due to the stabilization of aggregates by organic matter.

D. Studies in Pure Solutions

In order to study the effect of gypsum on the behavior of Al in the absence of complicating factors, it has been necessary to resort to the use of nutrient solutions. Alva et al. (1988), Noble et al. (1988b), and Tanaka et al. (1987) have presented compelling evidence to indicate that ion pair formation of $AlSO_4^+$ substantially detoxifies Al-containing solutions when gypsum is added, corroborating the original claim of Pavan et al. (1982). A further ameliorating effect is brought about by the presence of increased levels of Ca after gypsum application. Noble et al. (1988a) have proposed a Calcium–Aluminum Balance (CAB) parameter that best describes the phytotoxicity of a solution containing Al. When PG containing F and P in addition to SO_4 is added to a solution containing Al, Al–F complexes form that result in detoxification of the solution (Alva and Sumner, 1988). However the detoxification caused by F complex formation is unlikely to be of importance in the field, because most of the F in PG would be sorbed in the topsoil and would contribute little or nothing to subsoil amelioration. The same would be true for any P complexes that might form. Thus, the only complex likely to be of importance in subsoils is $AlSO_4^+$, provided an adequately high SO_4^{2-} concentration is maintained.

E. Column Leaching Studies

A number of leaching studies using both disturbed and undisturbed soil columns have been conducted to study, under controlled conditions, the effect of gypsum on the chemical behavior of soil profiles (Kotze and Deist, 1975; Lemus-Grob, 1985; Oates and Caldwell, 1985; Pavan et al. 1984; Reeve and Sumner, 1972; Sumner et al., 1985; O'Brien and Sumner, 1988). In all cases, the pattern of behavior has been essentially the same, and very similar to the results obtained under field conditions. Gypsum treatment results in increased levels of Ca and decreased levels of Al in the subsoil with little evidence for substantial movement of Al down the profile. Magnesium and sometimes K are removed in substantial quantities from the topsoil and may accumulate as transients in the subsoil before complete removal. Evidence from columns corroborates that obtained in the field, which suggests the immobilization of Al within the soil as some type of solid phase. In certain soils, there is strong evidence for the mobilization of Si when gypsum is added (Lemus-Grob, 1985; Sumner et al., 1985; O'Brien and Sumner, 1988), possibly as a result of the decomposition of kaolinite, which is less stable under acid conditions than some of the aluminum hydroxy sulfates.

F. Explanation of the Amelioration by Gypsum

Reeve and Sumner (1972) originally proposed a "self-liming" effect to account for the reduction in exchangeable Al in sesquioxidic soils after reaction with gypsum. According to this hypothesis, alkalinity from the surface exchange reaction:

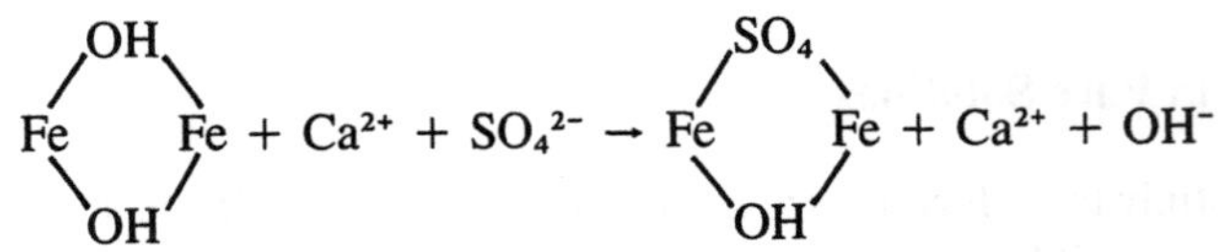

will precipitate some Al via reaction with OH^- released and reduce its toxicity. Most authors have since recognized that, while there is abundant evidence of a decrease in labile aluminum (expressed variously as exchangeable Al, Al saturation as a percentage of CEC, or the activity of the soluble Al^{3+} cation), usually the possibility remains that improved root development has resulted at least in part from an enhanced soil Ca status. Sumner et al. (1987) discussed the various mechanisms that might operate in the amelioration of acidic subsoils. The objective of this section is to summarize those arguments that currently offer the most satisfactory explanation for the beneficial effects of gypsum on root development which have been observed in acid soil. In recalling the various chemical effects of gypsum reaction with acid soil, a number of these can be discarded as primary explanations for improved root growth on the basis that they have not been consistently detected in all the experiments in which a positive root response was recorded. These include increases in soil solution pH per se (often but not always increased), in the solution activity of silicon, and the complexation of Al by the fluoride present as an impurity in by-product phosphogypsum (Alva and Sumner, 1988). By contrast, the features that emerge consistently are the increase in exchangeable Ca and the decrease (sometimes not very marked, but usually significant) in exchangeable Al. Generally the levels in soil solution of these metals follow the same pattern, although in the case of Al the effect is more complex in that the activity of Al^{3+} often decreases notwithstanding higher total Al concentration in solution, which sometimes follows treatment with gypsum (Table 18).

The relative effects of Ca and Al on root development in acidic subsoils are not easily distinguished, for these reasons:

1. The levels of soluble Ca and Al in soils are typically related inversely to one another, with leached, Ca-impoverished soils usually being sufficiently acidic to subtend toxic levels of Al in solution (exceptions are found in the form of extremely weathered Oxisols with such low CEC that Ca is deficient without Al necessarily being toxic) (Ritchey et al., 1980).
2. The toxic effects of Al are often difficult to interpret quantitatively, partly because it is the activity of Al rather than its concentration in solution that is most important (Foy, 1984). Also, soluble, monomeric Al appears to be less toxic when it occurs as ion pairs through hydrolysis or SO_4–complexation (Pavan et al., 1982; Noble et al. 1988a), F–complexation (Noble et al., 1988b), or complexation with organic anions (Hue et al., 1986).

3. The effects of Ca deficiency and Al toxicity on root development are not easily distinguished; there is evidence also that the supply of Ca helps lessen the toxic effects of a particular level of Al. This principle has been formulated in a number of expressions that were designed to characterize the acidity status of the soil (% Al or base saturation) or soil solution [Ca/Al concentration or activity ratio (Lund, 1970; Adams, 1984; Buyeye et al., 1985; Sumner et al., 1985; Noble et al., 1988a)] with respect to its effects on root growth.

In view of these considerations, there seems little point in attempting to force a distinction between enhanced Ca status and reduced Al toxicity in explaining the ameliorative effects of gypsum. Both are likely to be important in at least some soils. What is of particular interest, then, is the mechanism by which Al is rendered less labile by gypsum treatment, because simple cation-exchange considerations would suggest the opposite. In Figures 31 and 32, evidence for the "self-liming" effect (Reeve and Sumner, 1972) mentioned previously is to be found in the increase in pH (H_2O) and adsorbed sulfate (extractable with a phosphate solution) that parallels the decrease in exchangeable Al. The essential basis of this model is ligand exchange on sesquioxide surfaces, followed by hydrolysis and precipitation of exchangeable Al.

One alternative to this hypothesis is based on the possible precipitation of one or more basic Al sulfate minerals, the stability of which in acid soil environments enriched with sulfate is now well known (Adams and Rawajfih, 1977; Nordstrom, 1982). In interpreting the data of Pavan et al. (1982, 1984), Sposito (1985) suggested the treatment of a Brazilian Oxisol with gypsum might first give rise to the precipitation of a metastable basaluminite phase if the soil solution was sufficiently enriched with sulfate. This would then be followed by a slower conversion to a more stable mineral such as jurbanite. The extensive compilation of stability data by Nordstrom (1982) is especially useful for quantifying the reactions involved. It is found, for example, that for an approximately millimolar level of sulfate in solution, alunite will be more stable than kaolinite below a pH of about 4.5, with jurbanite only becoming the most stable Al mineral below pH 4. The consequences of a reaction of gypsum with aluminous minerals in soil should be identical to those predicted by the self-liming model. For example:

$$\underset{\text{(gibbsite)}}{Al(OH)_3} + CaSO_4 \rightarrow \underset{\text{(jurbanite)}}{AlOHSO_4} + Ca(OH)_2$$

$$\underset{\text{(kaolinite)}}{Al_2Si_2O_5(OH)_4} + 2CaSO_4 + 5H_2O \rightarrow \underset{\text{(jurbanite)}}{2AlOHSO_4} + 2H_4SiO_4 + 2Ca(OH)_2$$

Either of these reactions forms a basis for predicting a decrease in acidity after gypsum treatment, although the production of silicic acid in the reaction of gypsum with kaolinite might be expected to at least partially neutralize the $Ca(OH)_2$ produced, because the subsequent reaction that can be envisaged between silica and sesquioxides is known to be pH dependent (Bowden et al., 1980). Figure 36 summarizes the stability interrelationships among the aluminum minerals

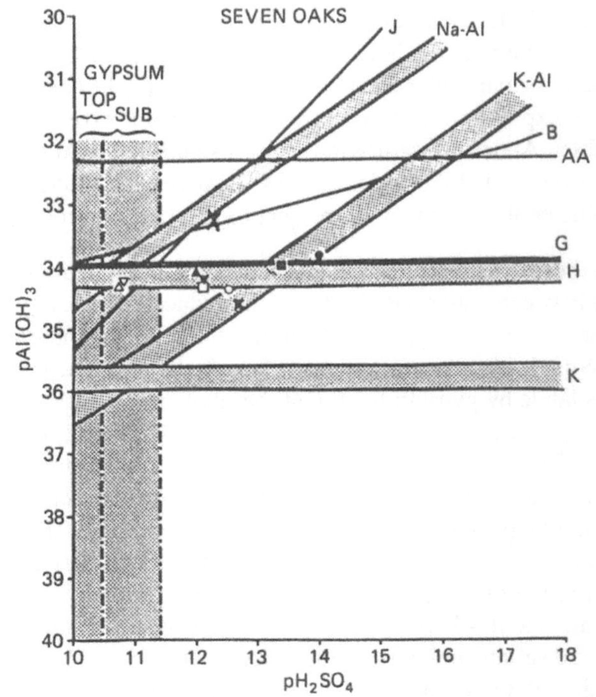

Figure 36. Equilibrium soil solution ion activity products [-log (Al^{3+}) $(OH^-)^3$ versus -log $(H^+)^2(SO_4{}^{2-})$] for South African soil treated with gypsum, KNO_3, or both, incubated for 2 months at field capacity. Points represent actual solution composition. (Topsoil: control, ○; KNO_3, □; gypsum, ▽; KNO_3 + gypsum, △. Subsoil: solid symbols, with corresponding meanings; ●, ■, ▼, and ▲, respectively.) Lines or bars represent solubility lines for the minerals indicated (J, jurbanite; Na-Al, natroalunite; K-Al, alunite; B, basaluminite; AA, amorphous $Al(OH)_3$; G, gibbsite; H, halloysite; K, kaolinite). Crosses represent solution composition of field subsoil sample (40–60 cm), either untreated (x) or treated (X) with 10 t ha^{-1} phosphogypsum 2 years previously (Fey, Noble, Buyeye, and Sumner, unpublished data).

involved, and provides examples of measured shifts in solution composition following gypsum treatment. Results such as these do not prove that the formation of one or more basic Al sulfate minerals has taken place; they merely serve as a basis for stating, on the grounds of supersaturation or undersaturation of the soil solution with respect to a particular mineral, whether or not the formation of such new minerals is thermodynamically feasible.

There appear to be some instances in which, after gypsum treatment, the soil solution is actually undersaturated with respect to all the solid phases depicted in Figure 36. This led Sumner et al. (1987) to suggest that surface reactions on sesquioxides, rather than mineral dissolution, may exert the ultimate control on solution composition. The nature of this control probably takes the form of a cosorption of Al^{3+} and SO_4^{2-}. Specific adsorption of sulfate imparts a more negative net charge to the surface (Bowden et al., 1980; Zhang et al., 1987). The resulting increase in negative surface charge density would cause a greater selectivity for Al^{3+} adsorption, the solution activity of which (and probably exchangeability by routine methods) would drop accordingly. In cases where an increase in solution pH is also measured, a similar process (in this case cosorption of protons) may have operated. To be tested quantitatively, this hypothesis would have to be developed further to include values for the constants in appropriately framed cation-exchange equilibrium reactions involving H^+, Al^{3+}, Ca^{2+}, and other dominant cations in the system.

A final aspect that needs to be addressed is the contention by Pavan et al. (1982, 1986) that Al toxicity is reduced by gypsum treatment primarily through ion pair formation. According to this hypothesis, Al remains in solution and may even increase in concentration, but the dominant species is now $AlSO_4^+$ (the data of Noble et al., 1988b, appear to confirm that the $AlSO_4^+$ species is less toxic to plant roots than Al^{3+}). Thermodynamically this hypothesis has less appeal, because the formation of $AlSO_4^+$ would be at the expense of Al^{3+}, and further dissolution of Al would therefore have to occur before equilibrium is reestablished. That this does happen is evident from some of the data in Table 18 [especially those of Buyeye et al. (1985); and even the 0.2–0.4 m Oxisol data of Pavan et al. (1984)]. In this respect, the mechanism of Al detoxification by gypsum addition to nutrient solutions with no solid phases present may be quite different from that which operates in acid soils. There is, of course, the possibility that the kinetics of Al dissolution to replenish that which is converted to $AlSO_4^+$ in solution are slow enough to make this proposed mechanism feasible in short-term amelioration.

The picture developed for the various reactions of Al in gypsum-amended acid soils is still too qualitative, and further clarification will probably only be attained when sufficient thermodynamic data (especially those pertaining to surface reactions involving cation- and ligand-exchange equilibria) become available for a more complete quantitative modeling of the system. What is apparent at this stage, however, is that, in explaining the detoxification of Al by gypsum addition, it makes little difference within which paradigm—adsorption or precipitation—one couches the explanation. Changes in pH and Al activity of the soil solution and in exchangeable (i.e., salt-extractable) Al in the whole soil are consistent with the "self-liming" reaction originally proposed by Reeve and Sumner (1972). This reaction can be predicted as a consequence of (1) ligand exchange of SO_4^{2-} for OH^- on sesquioxide surfaces; (2) Al^{3+} and H^+ adsorption in preference to Ca^{2+} on sesquioxide surfaces, which have been charged by the specific adsorption of SO_4^{2-}; and/or (3) a partial conversion of hydroxy Al minerals, in the presence of gypsum, into a basic Al sulfate solid and lime ($Ca(OH)_2$). In all these cases, the equilibrium activity of Al^{3+} (and therefore the potential for Al toxicity) must necessarily decrease. More often than not, the expected increase in soil solution pH is measurable. A reduction in exchangeable Al will follow.

Therefore, if surface application of amendments is to be employed, gypsum is uniquely suited to the amelioration of subsoil acidity. The sulfate radical is sufficiently mobile (unlike, say, phosphate) to penetrate the subsoil readily. Thereafter, its retention in a predominantly sesquioxidic medium is favored (unlike that of, say nitrate or chloride) by its propensity for bonding with Al (by specific adsorption and/or the formation of new solids). Calcium accompanies sulfate (an initial lag is usually observed because of the greater mobility of Mg) into the subsoil, where it rectifies a Ca deficiency and/or lessens the severity of Al toxicity, the latter having also been somewhat reduced by the self-liming effect of the sulfate–sesquioxide reaction. For these reasons, other salts of either Ca or SO_4 might be expected to produce markedly less ameliorative effects, although they may provide some benefit over untreated soils.

G. Practical Considerations

Notwithstanding the effectiveness of gypsum application as an acid subsoil ameliorant, any decision to adopt this strategy of combating subsoil acidity must necessarily be tempered by a number of practical considerations. The economics of gypsum use must be carefully considered in developing agriculture in particular. Distance from source is a factor that quite obviously has a marked effect on the viability of this approach. Per unit transportation costs are, in turn, directly influenced by the requirement per hectare, which appears to be high, perhaps prohibitively so, in some localities and soils. Rate of reaction is also slow in many situations and the farmer's ability to carry an extra debt load for two to three seasons cannot be ignored.

Unfortunately, few field trials have run long enough to facilitate reliable estimation of the time over which gypsum costs can reasonably be prorated. Other practical considerations that warrant consideration involve the method of application per se. Phosphogypsum is often difficult to spread because of its high moisture content, and there is clear evidence that the rate of gypsum movement into subsoil horizons is affected by tillage practices. Gypsum disked into previously deep-plowed soils, for example, moves far more rapidly than gypsum similarly incorporated on conventionally plowed soil. Does this mean that gypsum surface applied in minimum-till situations, a practice currently being tried by no-till farmers in parts of Brazil and South Africa, is likely to be significantly less effective than currently available information suggests?

Another quandary facing most producers in areas with acid soils is the need to rectify both surface and subsoil horizons. Gypsum is most certainly not an economic substitute for lime where topsoil acidity levels are excessive (in the South African Ultisol reported on here, two to three times the quantity of gypsum was required to achieve the same short-term effect on exchangeable Al and Al saturation in the topsoil as lime), and it is likely that many farmers would derive most benefit from conventionally liming of topsoils to desirable acidity levels. Only after this has been done should resources be allocated to subsoil problems.

Perhaps the most pertinent practical problem facing perspective users of gypsum is the lack of calibration data. There is currently no method of calculating gypsum requirement levels in different situations. From the findings discussed in previous sections, it is clear that requirements vary dramatically across localities and soils, even where essentially the same levels of acidity exist. Because overapplication of gypsum can have negative effects on crop growth, particularly on light-textured soils (Section III.H) and many soils have far less severe subsoil acidity problems than those studied thus far, farmers are faced with both economic and agronomic considerations of considerable importance. Much research is required before many needed answers can be obtained.

H. Adverse Effects of Phosphogypsum Use

Being a moderately soluble salt, gypsum applied at the surface is likely to replace other exchangeable cations, notably Mg and K, and render them prone to leaching. This is, in fact, the case illustrated in Table 20, where the effects of gypsum on topsoil Mg are followed with time. With increasing rates of application, the levels of exchangeable Mg are decreased in all three soils, and this tendency continues with time. The effects of gypsum on the redistribution of Mg and K in the Tifton profile with depth are presented in Table 21. The removal of Mg from the profile appears to be more efficient than that of K, as reflected by the total amounts of exchangeable Mg and K remaining in the profile. This depletion of Mg and K from the profile can have serious consequences, particularly on sandy soils such as the Tifton series.

On soils with greater cation-exchange capacities, this effect is likely to be much less, but nevertheless significant in terms of plant nutrition. Thus care should be exercised in the use of gypsum on sandy soils of low CEC. When gypsum is used on such soils, Mg and K application should be made to restore balanced nutrition after most of the gypsum has been leached from the topsoil.

Table 20. Effect of phosphogypsum on topsoil-exchangeable magnesium with time

	Tifton		Davidson		Altavista	
Treatment	1985	1986	1985	1986	1985	1986
Control	0.25[a]	0.30	1.31	1.22	0.33	0.23
PG (2 Mg ha^{-1})	0.13	0.13	0.92	0.82	0.27	0.22
PG (5 Mg ha^{-1})	0.09	0.06	0.95	0.67	0.23	0.20
PG (10 Mg ha^{-1})	0.15	0.06	0.47	0.27	0.18	0.16
Mixed and limed to 1 m deep	0.29	0.30	1.01	1.15	0.38	0.31
$LSD_{0.05}$		0.04		0.18		0.06

From Syed (1977).

[a] All values in cmol(+) kg^{-1}.

Table 21. Effect of phosphogypsum on exchangeable Mg and K down the profile of a Tifton series soil

Depth (m)	Phosphogypsum rate			
	0 Mg ha^{-1}		10 Mg ha^{-1}	
	Mg	K	Mg	K
0–0.15	0.25	0.12	0.15	0.10
0.15–0.30	0.23	0.14	0.10	0.10
0.30–0.45	0.10	0.07	0.04	0.05
0.45–0.60	0.60	0.06	0.03	0.04
0.60–0.75	0.08	0.08	0.04	0.05
0.75–0.90	0.09	0.07	0.05	0.06
0.90–1.05	0.10	0.08	0.09	0.06
Total	0.91	0.62	0.50	0.46

Note. All values in cmol(+) kg^{-1}.
From Syed (1987).

I. Gypsum As a Nutrient Source

Gypsum has been used as a fertilizer material since the latter part of the eighteenth century (Crocker, 1945). Benjamin Franklin, after returning from Europe, is reputed to have demonstrated its value to American farmers by spreading gypsum on his clover field in a pattern outlining the words "This land has been plastered." In retrospect, it seems probable that the improved growth resulted from enhanced S nutrition, the ready availability of S in gypsum having been documented by workers too numerous to cite. There are indications that in certain soils the mobility of SO_4^{2-} from gypsum is such that less readily available forms of S may be required for longer term benefits (McLachlan and DeMarco, 1968). Conversely, sulfate is regarded as a superior S source in submerged crops, as the oxidation of elemental S to the sulfate form taken up by crops may be retarded (Wang, 1978). It also seems likely that gypsum will be superior to other elemental forms of S or ammonium sulfate in situations in which further acidification is undesirable. This would apply to soils that are naturally strongly acidic, a common situation on cultivated soils in the tropics and subtropics.

There is little doubt that S nutritional problems are likely to increase worldwide with the move to higher grade fertilizers, mandatory reductions in SO_2 emissions, and increased S removal because of higher crop yields, but future gypsum use patterns are less clear. Transportation and handling costs have been primarily responsible for the move to higher grade fertilizers, and it is probable that in broad terms gypsum will only remain competitive with higher grade S fertilizers where the advantages already mentioned are considered important or where the land requiring gypsum is reasonably close to a source. Palmer et al. (1983) have discussed a range of alternative S sources of higher analysis than gypsum, and combinations of triple superphosphate and elemental S have been used successfully on high-S-requiring pastures in New Zealand and elsewhere (Sinclair and Saunders, 1982).

The special place gypsum has enjoyed as a Ca fertilizer material for peanuts is well known (Mehlich and Tewari, 1974), and it is unlikely that its role in this regard will be unduly affected by transport and handling considerations. It has been clearly demonstrated (Smal et al., 1988; Walker and Csinos, 1980; Walker et al., 1979) that adequate quantities of Ca must be present in the pegging zone for the proper development of disease-free peanuts. Gypsum has been the most usual source. In fact, the possibility exists that appreciation of gypsum's role as a Ca source may be extended to other crops. In chemically equivalent terms, the quantity of gypsum required to increase the activity of Ca in the soil solution is very much less than that of lime (Figure 28), and there is evidence to suggest that adequate Ca levels are difficult to subtend with lime in strongly acidic, variable charge soils (Fox et al., 1985). The quality of several crops other than peanuts is very dependent on Ca supply. Gypsum has, for example, been shown to decrease storage rots of potatoes (McGuire and Kelman, 1984) and to reduce the incidence of bitter pit in apples (van der Boon, 1980). In the case of potatoes, the deleterious effects of lime on the incidence of scab have led to widespread use of gypsum in parts of South Africa where potatoes are produced on acidic soils that are low in Ca. In Australia gypsum is also reputed to be more effective than lime in combating the effects of *Phytophthora cinnamoni* in avocados (Trochoulias et al., 1986).

IV. Environmental Concerns in Phosphogypsum Use

Potential environmental concerns are an important consideration in suggesting any material as a soil amendment, particularly a waste or by-product such as phosphogypsum or other industrial-type gypsums. While the major constituents (Ca and SO_4) are not likely to be considered environmentally damaging in ground or surface waters, waste gypsums may contain trace elements, particularly metals, carried over from the primary processing.

During the manufacture of phospho- and fluorogypsum, ^{226}Ra is retained in the by-product gypsum, whereas other radioactive elements (^{238}V, ^{232}Th) contained in the phosphate rock are removed in the phosphoric acid. Although there are no federal regulations limiting the levels of radioactivity in PG in terms of its storage and uses, the effect of the ^{226}Ra content on soils and crops has been examined by Mays and Mortvedt (1986). Using rates much higher than would normally be used under field conditions, they found that radioactivity levels in the grain of three crops were not affected by PG applications (Table 22). However, PG at 112 Mg ha^{-1} did double the level of ^{226}Ra radioactivity in wheat grain, but Mays and Mortvedt (1986) considered this to be relatively unimportant as the differential ^{226}Ra levels in the soil were not manifest in crop uptake. Rock phosphate, particularly from Florida, often also contains substantial quantities of Cd.

Table 22. Effect of phosphogypsum on Cd and radioactivity levels in corn, wheat, and soybean grain

Rate of phosphogypsum ($Mg\ ha^{-1}$)	Cd[a]			Radioactivity[b]		
	Corn	Wheat	Soybean	Corn	Wheat	Soybean
0	0.03	0.05	0.05	0.04	0.46	0.84
22	0.04	0.04	0.06	0.03	0.37	0.37
112	0.03	0.06	0.07	0.03	0.83	0.37

After Mays and Mortvedt (1986).

[a] Data are in $mg\ kg^{-1}$.

[b] Data are in $Bq\ kg^{-1}$. Counting error (1 SD) was 0.02, 0.37, and 0.37 $Bq\ kg^{-1}$ for corn, wheat, and soybeans, respectively.

However, in terms of the content in grain, this Cd appears to be of no concern environmentally (Table 22), being only a fraction of values reported in situations where Cd had been added to soil as municipal sludges or as phosphatic fertilizers (Giordano et al., 1975; Mortvedt et al., 1981). Recoveries in grain of Cd and radioactivity from soil treated with PG often are of the same magnitude or less than in the control (Table 23) and are always a very small fraction of that added to soil. Thus, PG should present no environmental problems when used at appropriate rates for agricultural purposes (0–10 $Mg\ ha^{-1}$).

Gypsum produced from desulfurization of coal stack gases will be likely to contain volatile metals (As and Se) as well as variable amounts of fly ash not previously removed from the gas stream (Behrens and Hargrove, 1980). Composition will likely be variable, and loading rates and mobility of added contaminants will have to be monitored on soils amended with these products.

A positive environmental impact of gypsum use on agricultural soils is the observed decrease in total sediment and in clay-sized sediments generated from gypsum-amended soils (Miller and Scifres, 1988). Flocculation of eroded clay caused by the presence of dissolved gypsum should decrease the transport of sediment into surface waters, thereby reducing the amount of sediment-associated agricultural chemicals leaving the field. This effect may be transitory, as deposited clay may be redispersed on exposure to low ionic strength waters; but combined with other sediment-control practices (terraces, sediment basins, grassed waterways), sedimented clay may be stabilized on or near the field with less chance of later remobilization.

As a final consideration, by-product gypsum often poses an environmental problem adjacent to processing facilities where concentrated leachate from large stockpiles may pose a threat to water supplies; the large "stacks" themselves (some near the Florida phosphate fertilizer works are more than 40 m tall) pose an aesthetic problem (May and Sweeney, 1983). Agronomic use of this material is highly desirable from the point of view of eliminating the need for stockpiling.

Table 23. Recovery of Cd and radioactivity added to soil in phosphogypsum by corn, wheat, and soybean grain

Rate of phosphogypsum ($Mg\ ha^{-1}$)	Cd added to soil ($g\ ha^{-1}$)	Cd removal in grain ($g\ ha^{-1}$)			Radioactivity added to soil ($Bq\ ha^{-1}$)[a]	Radioactivity in grain ($Bq\ ha^{-1}$)[a]		
		Corn	Wheat	Soybeans		Corn	Wheat	Soybeans
0	0	0.41	0.17	0.15	0	548	1578	2587
22	5.06	0.53	0.15	0.18	20.3×10^6	399	1417	1117
112	25.76	0.33	0.23	0.20	103.6×10^6	330	3120	1106

Calculated from data of Mays and Mortvedt (1986).

[a] 1 Bq = 1 disintegration per second.

V. General Conclusions

The major objective of this review is to show that gypsum, a widely available and generally inexpensive material, has a wide range of agronomic applications because of its favorable effect on both soil physical properties (on dispersive soils) and chemical properties (on acid soils). Some of these effects are well documented in the literature, while others are under active current investigation. Although present gypsum use worldwide is limited to reclamation of sodic soils or as a nutrient source for Ca or S, current research is suggesting a much more widespread potential for use on dispersive and/or highly acid soils. On crusting soils, particularly in semiarid or marginal rainfall areas, gypsum may significantly increase water availability, emergence, or both, and prove to be at least a short-term solution to a serious limitation in production. On acid soils, the improvements in subsoil fertility with gypsum amendment, and subsequent yield increases, offer a significant new approach to managing these soils, which often show poor or variable yields as the result of restricted rooting. The combination of effects on some soils may improve yields and production efficiency synergistically by allowing more water to infiltrate and encouraging roots to exploit water that is stored in subsoils horizons.

The expanded use of gypsum will depend in large part on the economics of production, which vary spatially and temporally. The local availability of gypsum and its transport cost, the probability and magnitude of yield increases, the optimum rate and need for reapplication, and the availability of capital that can be amortized over several growing seasons will be important factors affecting profitability, and will probably vary on different soil–crop combinations. In industrial countries with supplies of cheap by-product gypsum, intensive agronomic research programs, and readily available credit, gypsum use may well expand, especially on Ultisols where crusting and subsoil acidity are both yield limiting. In developing countries, these factors may work against gypsum use, despite the probable yield increases to be obtained in both semiarid and tropical region soils.

As is obvious from this discussion, further research is needed in many areas to clarify the processes and effects associated with use of gypsum on soils. With respect to both effects on water relations and subsoil amelioration, the following may be suggested as areas for further investigation:

1. Long-term variable-rate studies across a wide range of soils are required to establish expected reaction rates, optimal levels of application, and the longevity of beneficial effects. Such information is critical to any assessment of the economic benefits of gypsum use in specific situations and to comparison with alternate strategies of either soil crusting reduction, subsoil amelioration, or both.
2. Information is required on the effects of incorporation depth and soil properties on the rate of vertical movement of added gypsum. Particularly important here are the effects of gypsum in no-till systems. In the humid tropics such systems are considered critical for soil protection (e.g., more than 1,000,000 ha of no-till in Brazil).

3. Fundamental information on soil dispersion and its relationship to soil structural stability and infiltration is needed for modeling the action of chemical parameters such as electrolyte concentration on the physical behavior of various soils, particularly those of the humid regions.
4. An explanation should be found for the apparent depressive effect of gypsum on root proliferation of some crops in the zone of maximum sulfate accumulation. Long-term studies on variable rates and frequency of application would provide suitable vehicles for such investigations.
5. More direct comparisons between the strategies available for combating the effects of subsoil acidity are needed. Maximum benefits could possibly be obtained from combining two or more available options.
6. The chemistry of gypsum reactions with acid soil is still imperfectly understood, especially with reference to the magnitude of its effect on labile aluminum. A rigorous thermodynamic model that predicts partitioning into solid and solution phases is probably needed before a quantitative description of all the reactions taking place can be achieved. Such understanding is probably necessary to develop a basis for predicting soils on which gypsum will be effective and how much should be applied to optimize the effect.

Whether continuing studies such as those suggested result in more widespread use of gypsum in agriculture is only part of the importance of current and future studies in this area: Much insight has been gained on basic soil processes, such as dispersion and Ca/Al interactions, and this will contribute to our overall understanding of the soil system and will undoubtedly improve our ability to manage it in the future.

Acknowledgment

Preparation of this review was supported by grants from the Florida Institute of Phosphate Research, and the South African Council for Scientific and Industrial Research, and the South African Department of Agriculture and Water Supply.

References

Abruna, F., J. Vicente-Chandler, and R.W. Pearson. 1964. Effects of liming on yields and composition of heavily fertilized grasses and on soil properties under humid tropical conditions. *Soil Sci. Soc. Am. Proc.* 28:657–661.

Adams, F. 1984. *Soil Acidity and Liming*, 2d Ed. American Society of Agronomy, Madison, Wisconsin.

Adams, F., and B.L. Moore. 1983. Chemical factors affecting root growth in subsoil horizons of Coastal Plain soils. *Soil Sci. Soc. Am. J.* 47:99–102.

Adams, F., and Z. Rawajfih. 1977. Basaluminite and alunite: A possible cause of sulfate retention by acid soils. *Soil Sci. Soc. Am. J.* 41:686–692.

Adeoye, K.B., and L. Singh. 1985. The effect of bulk application of lime under two tillage depths on soil pH and crop yield. *Plant Soil* 85:295–298.

Agassi, M., J. Morin, and I. Shainberg. 1982. Infiltration and runoff control in the semi-arid region of Israel. *Geoderma* 28:345–355.

Agassi, M., J. Morin, and I. Shainberg. 1985a. The effect of water drop impact energy and water salinity on the infiltration rate of sodic soils. *Soil Sci. Soc. Am. J.* 49:186–190.

Agassi, M., I. Shainberg, and J. Morin. 1985b. Infiltration and runoff in wheat fields in the semi-arid region of Israel. *Geoderma* 36:263–276.

Agassi, M., I. Shainberg, and J. Morin. 1981. Effect of electrolyte concentration and soil sodicity on infiltration rate and crust formation. *Soil Sci. Soc. Am. J. 45:848*–851.

Agassi, M., I. Shainberg, and J. Morin. 1986. Effect of powdered phosphogypsum on the infiltration rate of sodic soils. *Irrig. Sci.* 7:53–61.

Alva, A.K., and M.E. Sumner. 1988. Effects of phosphogypsum and calcium sulfate on aluminum speciation in solutions and soybean root growth. *Soil Sci. Soc. Am. J.* 52 (in press).

Alva, A.K., M.E. Sumner, and A.K. Noble. 1988. Alleviation of aluminum toxicity by phosphogypsum. *Commun. Soil Sci. Plant Anal.* 19 (in press).

Anderson, D.L., and J.G. Hendrick. 1983. Subsoil lime injector. *Soil Sci. Soc. Am. J.* 47:337–339.

Anonymous. 1986. *Anais do e Semiario Sobre o Uso do Fosfogesso na Agricultura.* EMBRAPA-DDT, Brasilia DF, Brazil.

Arora, H.S., and N.T. Coleman, 1979. The influence of electrolyte concentrations on flocculation of clay suspensions. *Soil Sci.* 127:134–139.

Aylmore, L.A.G., and J.P. Quirk. 1959. Swelling of clay–water systems. *Nature (Lond)* 183:1752–1753.

Banin, A., and N. Lahav. 1968. Particle size and optical properties of montmorillonite in suspension. *Isr. J. Chem.* 6:235–250.

Bar-on, P., I. Shainberg, and I. Michaeli. 1970. The electrophoretic mobility of Na/Ca montmorillonite particles. *J. Colloid Interface Sci.* 33:471–472.

Barton, F.M., and N.M. Wilde. 1979. Dissolution rates for polycrystalline samples of gypsum and orthorombic forms of calcium sulfate, by a rotating disk method. *Faraday Soc. Trans* 67:3590–3597.

Bashir, A., W.D. Kemper, Ghulam Haider, and M.A. Niazi. 1979. Use of gypsum stones to lower the sodium adsorption ratio of irrigation water. *Soil Sci. Soc. Am. J.* 43:698–702.

Baver, L.D., W.H. Gardner, and W.R. Gardner. 1972. *Soil Physics*, 4th Ed. Wiley, New York.

Behrens, G.P., and O.W. Hargrove. 1980. *Evaluation of Chiyoda Thoroughbred 121 FGD Process and Gypsum Stacking*. EPRI Project 536–4, Electric Power Research Institute, Palo Alto, California.

Ben-Hur, M., I. Shainberg, and J. Morin. 1987. Variability of infiltration in a field with surface-sealed soil. *Soil Sci. Soc. Am. J.* 51:1299–1302.

Bennett, O.L., D.A. Ashley, and B.D. Doss. 1964. Methods of reducing soil crusting to increase cotton seedling emergence. *Agron. J.* 56:162–165.

Bernstein, L. 1974. Crop growth and salinity. In: J. Van Schilfgaarde (ed.), *Drainage for Agriculture*. American Society of Agronomy, Madison, Wisconsin.

Blackmore, A.V., and R.D. Miller. 1961. Tactoid size and osmotic swelling in Ca montmorillonite. *Soil Sci. Soc. Am. Proc.* 25:169–173.

Bouldin, D.R. 1979. The influence of subsoil acidity on crop yield potential. *Cornell Int. Agric. Bull 34.*

Bouton, J.H., and M.E. Sumner. 1983. Alfalfa, *Medicago sativa* L., in highly weathered acid soils. V. Field performance of alfalfa selected for acid tolerance. *Plant Soil* 74:430–436.

Bowden, J.W., A.M. Posner, and J.P. Quirk. 1980. Adsorption and charging phenomena in variable charge soils. In: B. Theng (Ed.), *Soils with Variable Charge*. New Zealand Soil Science Society, Lower Hutt, New Zealand.

Bresler, E. 1970. Numerical solution of the equation for interacting diffuse layer in mixed ionic system with non-symmetrical electrolytes. *J. Colloid Interface Sci.* 3:278–283.

Bresler, E., B.L. McNeal, and D.L. Carter. 1982. *Saline and Sodic Soils*. Springer-Verlag, New York.

Brown, B.A., and R.I. Munsell. 1938. Soil acidity at various depths as influenced by the time since application, placement and amount of limestone. *Soil Sci. Soc. Am. Proc.* 3:217–221.

Buyeye, S.M., and M.V. Fey. 1987. Several explanations for the amelioration of acid, sesquioxidic soils by gypsum application. *Proc. Soil Sci. Soc. S. Afr.* (in press).

Buyeye, S.M., M.V. Fey, and C.J.B. Mott. 1985. Maize growth and acid soil responses to treatment with gypsum and lime. *Proceedings of the South African Society Crop Production Congress* pp. 366–384. South African Society of Crop Production, Coetzenberg, South Africa.

Carvalho, L.J.C.B., R.L. Gomide, G.C. Rodrigues, D.M.G. Souza, and E. de Freitos, Jr. 1986. Reposta do Milho à Aplicação de Gesso e Déficit Hidrico em Solo de Cerrados. *An. Sem. Uso. Fosfogesso Agricultura,* pp. 61–83, Brasilia D.F., Brazil.

Cary, J.W., and D.D. Evans. 1974. Soil crusts, *Tech. Bull. No. 214*, University of Arizona, Tucson, Arizona.

Chaudri, K.G., and D.K. Das. 1977. Physical characteristics of soil crusts. *J. Soc. Exp. Agric.* 2:40–45.

Chaudhry, G.H., and B.P. Warkentin. 1968. Studies on exchange of sodium from soils by leaching with calcium sulphate. *Soil Sci.* 105:190–197.

Chen, Y., J. Tarchitzky, J. Morin, and A. Banin. 1980. Scanning electron microscope observations on soil crusts and their formation. *Soil Sci.* 130:49–55.

Chiang, S.C., D.E. Radcliffe, W.P. Miller, and K.D. Newman. 1987. Hydraulic conductivity of three southeastern soils as affected by sodium, electrolyte concentration and pH. *Soil Sci. Soc. Am. J.* 51:1293–1299.

Cole, H.H. 1930. The gypsum industry in Canada. *Canada Dept. of Mines, Mines Branch Rep. No. 714.* F.A. Acland, Ottawa, Canada.

Couto, W., D.J. Lathwell, and D.R. Bouldin. 1979. Sulfate sorption by two Oxisols and an Alfisol of the tropics. *Soil Sci.* 127:108–116.

Crocker, W. 1945. Sulfur deficiency in soils. *Soil Sci.* 60:149–155.

Dematte, J.L.I. 1981. Characteristics of Brazilian soils related to root growth. In: R.S. Russell, K. Igue, and Y.R. Mekta (eds.), *The Soil/Root System in Relation to Brazilian Agriculture*, pp. 21–41. Fundacao Instituto Agronomico do Parana, Brazil.

Doss, B.D., W.T. Dumas, and Z.F. Lund. 1979. Depth of lime incorporation for correction of subsoil acidity. *Agron. J.* 71:541–544.

Doner, H.E., and W.C. Lynn. 1977. Carbonate, halide, sulfate, and sulfide minerals. pp. 75–98. In: J.B. Dixon and S.B. Weed (eds.), *Minerals in Soil Environments.* Soil Science Society of America, Madison, Wisconsin.

Duley, F.L. 1939. Surface factors affecting the rate of intake of water by soils. *Soil Sci. Soc. Am. Proc.* 4:60–64.

Durgin, P.B., and J.G. Chaney. 1984. Dispersion of kaolinite by dissolved organic matter from Douglas-Fir roots. *Can. J. Soil Sci.* 64:445–455.

Dutt, G.R., and R.W. Terkeltoub. 1972. Prediction of gypsum and leaching requirements for sodium-affected soils. *Soil Sci.* 114:93–103.

Elgabaly, M.M., and W.M. Elghamry. 1970. Water permeability and stability of kaolinite systems as influenced by adsorbed cation ratio. *Soil Sci.* 110:107–110.

Ellison, W.D. 1947. Soil erosion studies. IV. Soil detachment by surface flow. *Agric. Eng.* 28:442–444, 450.

El-Swaify, S.A., and L.D. Swindale. 1969. Hydraulic conductivity of some tropical soils as a guide to irrigation water quality. *Int. Congr. Soil Sci., Trans. 9th* 1:381–389.

El-Swaify, S.A., and E.W. Dangler. 1977. Erodibility of selected tropical soils in relation to structural and hydrological parameters. In: G.R. Foster (Ed.), *Soil Erosion: Prediction and Control.* Soil Conservation Society of America, Iowa.

Emerson, W.W. 1967. A classification of soil aggregates based on their coherence in water. *Aust. J. Soil Res.* 5:47–57.

Evans, D.D., and S.W. Buol. 1968. Micromorphological study of soil crust. *Soil Sci. Soc. Am. Proc.* 32:19–22.

Farina, M.P.W., and P. Channon. 1988a. Acid-subsoil amelioration I. A comparison of several mechanical procedures. *Soil Sci. Soc. Am. J.* 52:169–174.

Farina, M.P.W., and P. Channon. 1988b. Acid-subsoil amelioration II. Gypsum effects on growth and subsoil chemical properties. *Soil Sci. Soc. Am. J.* 52:175–180.

Felhendler, R., I. Shainberg, and H. Frenkel. 1974. Dispersion and hydraulic conductivity of soils in mixed solution. *Trans. 10th Int. Congr. Soil Sci. (Moscow)* I:103–112. Nauka, Moscow.

Foster, G.R. 1982. Modelling the erosion process. In: C.T. Haan (Ed.), *Hydrologic Modelling of Small Watersheds*, pp. 297–379. American Society of Agricultural Engineers, St. Joseph, Missouri.

Fox, R.L., R.S. Yost, N.A. Saidy, and B.T. Kang. 1985. Nutritional complexities associated with pH variables in humid tropical soils. *Soil Sci. Soc. Am. J.* 49:1475–1480.

Foy, C.D. 1974. Effects of aluminum on plant growth. In: E.W. Carson (ed.), *The Plant Root and Its Environment*, p. 601–642. University Press of Virginia, Charlottesville, Virginia.

Foy, C.D. 1984. Physiological effects of hydrogen, aluminum, and manganese toxicities in acid soils. In: F. Adams (ed.), *Soil Acidity and Liming*, 2nd Ed., pp. 57–98. American Society of Agronomy, Madison, Wisconsin.

Frenkel, H., J.O. Goertzen, and J.D. Rhoades. 1978. Effects of clay type and content, exchangeable sodium percentage, and electrolyte concentration on clay dispersion and soil hydraulic conductivity. *Soil Sci. Soc. Am. J.* 42:32–39.

Frenkel, H., and A. Hadas. 1981. Effect of tillage and gypsum incorporation on rain runoff and crust strength in field soils irrigated with saline-sodic water. *Soil Sci. Soc. Am. J.* 45:156–158.

Gal, M., L. Arcan, I. Shainberg, and R. Keren. 1984. The effect of exchangeable Na and phosphogypsum on the structure of soil crust—SEM observations. *Soil Sci. Soc. Am. J.* 48:872–878.

Giordano, P.M., J.J. Mortvedt, and D.A. Mays. 1975. Effect of municipal wastes on crop yield and uptake of heavy metals. *J. Environ. Qual.* 4:394–399.

Goedert, W.J. 1983. Management of the Cerrado soils of Brazil: A review. *J. Soil Sci.* 34:405–428.

Goldberg, S., and R.A. Glaubig. 1987. Effect of saturating cation, pH, and aluminum and iron oxide on the flocculation of kaolinite and montmorillonite. *Clays Clay Miner.* 35:220–227.

Gonzalez-Erico, E., E.J. Kamprath, G.C. Nademan, and W.V. Soares. 1979. Effect of depth of lime incorporation on the growth of corn on an Oxisol of Central Brazil. *Soil Sci. Soc. Am. J.* 43:1155–1158.

Guimarães, P.T.C. 1986. O. Gesso Agrícola n Neutralização do Aluminio Nas Camadas Subsuperficias do Solo: Aplicações às Cultivars Anuais e Perenes, pp. 145–168. In: *An Sem. Uso Fosfogesso Agricultura,* EMBRAPA, Brasilia D.F., Brazil.

Hamblin, A., and M. Howell. 1988. Maintenance and improvement of soil structure. *Tech. Bull.*, Western Australia Dept. of Agriculture. (in press).

Hammel, J., M.E. Sumner, and H. Shahandeh. 1985. Effect of physical and chemical profile modification on soybean and corn production. *Soil Sci. Soc. Am. J.* 49:1508–1512.

Hardy, N., I. Shainberg, M. Gal, and R. Keren. 1983. The effect of water quality and storm sequence upon infiltration rate and crust formation. *J. Soil Sci.* 34:665–676.

Haynes, R.J. 1984. Lime and phosphate in the soil–plant system. *Adv. Agron.* 37:249–315.

Hilgard, E.W. 1906. *Soils—Their Formation, Properties, Composition and Relation to Climate and Plant Growth in the Humid and Arid Regions* Macmillan, London.

Hillel, D. 1980. *Applications of Soil Physics.* Academic Press, New York.

Howell, M. 1987. Gypsum use in the wheat belt. *J. Agric. W. Aus.* 28:40–43.

Hue, N.V., F. Adams, and C.E. Evans. 1985. Sulfate retention by an acid BE horizon of an Ultisol. *Soil Sci. Soc. Am. J.* 49:1196–1200.

Hue, N.V., G.R. Craddock, and F. Adams. 1986. Effect of organic acids on aluminum toxicity in subsoils. *Soil Sci. Soc. Am. J.* 50:28–34.

Huett, D.O., and R.C. Menery. 1980. Effect of aluminum on growth and nutrient uptake of cabbage, lettuce and kikuyu grass in nutrient solution. *Aust. J. Agric. Res.* 31:749–761.

Hurlbut, C.S., and C. Klein. 1971. *Manual of Mineralogy*, 19th Ed. Wiley, New York.

Kauffman, M.D., and E.H. Gardner. 1978. Segmental liming of soil and its effect on the growth of wheat. *Agron. J.* 70:331–336.

Kazman, Z., I. Shainberg, and M. Gal. 1983. Effect of low levels of exchangeable Na and applied phosphogypsum on the infiltration rate of various soils. *Soil Sci.* 135:184–192.

Keisling, T.C., P.S.C. Rao, and R.E. Jessup. 1978. Pertinent criteria for describing the dissolution of gypsum beds in flowing water. *Soil Sci. Soc. Am. J.* 42:234–246.

Kelley, W.P., and A. Arany. 1928. The chemical effect of gypsum, sulphur, iron sulphate and alum on alkali soil. *Hilgardia* 3:393–420.

Kemper, W.D., J. Olsen, and C.J. DeMooy. 1975. Dissolution rate of gypsum in flowing water. *Soil Sci. Soc. Am. Proc.* 39:458–463.

Keren, R., and I. Shainberg. 1981. Effect of dissolution rate on the efficiency of industrial and mined gypsum in improving infiltration of a sodic soil. *Soil Sci. Soc. Am. J.* 45:103–107.

Keren, R., I. Shainberg, H. Frenkel, and Y. Kalo. 1983. A field study of the effect of exchangeable sodium on surface runoff from loess soil. *Soil Sci. Soc. Am. J.* 47:1001–1004.

Keren, R., I. Shainberg, and E. Klein. 1988. Settling and flocculation value of sodium-montmorillonite particles in aqueous media. *Soil Sci. Soc. Am. J.* 52:76–80.

Kiehl, J.C., and O. Franco. 1984. Efecto do Gesso Industrial Solve Alguns Componentes da Fertilidade do Solo. *O. Solo* 76:11–16.

Kotze, W.A.G., and J. Deist. 1975. Amelioration of subsurface acidity by leaching of surface applied amendments. A laboratory study. *Agrochemophysica* 7:39–46.

Lal, R. 1983. Effects of slope length on runoff from Alfisols in western Nigeria. *Geoderma* 31:185–193.

Lemus-Grob, F. 1985. *Effects and Behavior of Gypsum in an Ultisol.* M.S. Thesis, University of Georgia, Athens, Georgia.

Loveday, J. 1974. Recognition of gypsum-responsive soils. *Aust. J. Soil Res.* 12:87–96.

Loveday, J. 1976. Relative significance of electrolyte and cation exchange effects when gypsum is applied to a sodic clay soil. *Aust. J. Soil Res.* 14:362–371.

Loveday, J. 1984. Amendments for reclaiming sodic soils. In: I. Shainberg, and J. Shalhevet (eds.), *Soil Salinity under Irrigation*, pp. 220–237. Springer-Verlag, New York.

Loveday, J., and J.C. Pyle. 1973. The Emersion dispersion test and its relationship to hydraulic conductivity. *CSIRO Aust. Div. Soils Tech. Paper 15.*

Loveday, J., and D.R. Scotter. 1966. Emergence response of subterranean clover to dissolved gypsum in relation to soil properties and evaporative conditions. *Aust. J. Soil Res.* 49:55–68.

Lund, Z.F. 1970. The effect of calcium and its relation to several cations in soybean root growth. *Soil Sci. Soc. Am. Proc.* 34:456–459.

Malavolta, E., M.R. Guilherme, and T.H. Liem. 1986. Associações fosfogesso calčario:principios e aplicações. pp. 177–196. In: *I Seminario Sobre o Uso do Fosfogesso na Agricultura*. EMBRAPA, Brasilia, Brazil.

Manin, M., A. Pissarra, and J.W. Van Hoorn. 1982. Drainage and desalinization of heavy clay soil in Portugal. *Agric. Water Manage.* 5:227–240.

May, A., and J.W. Sweeney. 1983. Evaluation of radium and toxic element leaching characteristics of Florida phosphogypsum stockpiles. *Rep. Investigations/Bureau of Mines No. 8776.* U.S. Govt. Printing Office, Washington, D.C.

Mays, D.A., and J.J. Mortvedt. 1986. Crop response to soil applications of phosphogypsum. *J. Environ. Qual.* 15:78–81.

McAtee, J.L., 1961. Heterogeneity in montmorillonites. *Clays Clay Miner.* 5:279–288.

McGuire, R.G., and A. Kelman. 1984. Reduced severity of Erwinia soft rot in potato tubers with increased calcium content. *Phytopathology* 74:1250–1256.

McIntyre, D.S. 1958. Permeability measurements of soil crusts formed by raindrop impact. *Soil Sci.* 85:185–189.

McIntyre, D.S. 1979. Exchangeable sodium, subplasticity and hydraulic conductivity of some Australian soils. *Aust. J. Soil Res.* 17:115–120.

McIntyre, D.S., J. Loveday, and C.L. Watson. 1982. Field studies of water and salt movement in an irrigated swelling clay soil. I. Infiltration during ponding. II. Profile hydrology during ponding. III. Salt movement during ponding. *Aust. J. Soil Res.* 20:81–90, 90–99, 101–105.

McKenzie, R.C., and M. Nyborg. 1984. Influence of subsoil acidity on root development and crop growth in soils of Alberta and northeastern British Columbia. *Can. J. Soil Sci.* 64:681–697.

McLachlan, K.D., and D.G. De Marco. 1968. The influence of time of application of gypsum and elemental sulphur on the pasture response to sulphur. *Aust. J. Exp. Agric. Anim. Husb.* 8:725–730.

McNeal, B.L., and N.T. Coleman. 1966. Effect of solution composition on soil hydraulic conductivity. *Soil Sci. Soc. Am. Proc.* 30:308–312.

McNeal, B.L., D.A. Layfield, W.A. Norvell, and J.D. Rhoades. 1968. Factors influencing hydraulic conductivity of soils in the presence of mixed-salt solutions. *Soil Sci. Soc. Am. Proc.* 32:187–190.

Mehlich, A., and G.P. Tewari. 1974. *The roles of gypsum (calcium sulphate) in agriculture*. U.S. Gypsum Co., Chicago, Illinois.

Mering, J., and R. Glaeser. 1954. On the role of the valency of exchangeable cations in montmorillonite. *Bull. Soc. Fr. Miner. Cristallogr.* 77:519–530.

Middleton, H.E. 1930. Properties of soils which influence soil erosion. *U.S. Dept. Agric. Tech. Bull. 178*, 16 pp.

Miller, W.P. 1987a. Infiltration and soil loss of three gypsum-amended Ultisols under simulated rainfall. *Soil Sci. Soc. Am. J.* 51:1314–1320.

Miller, W.P. 1987b. A solenoid-operated variable intensity rainfall simulator. *Soil Sci. Soc. Am. J.* 51:832–834.

Miller, W.P., and M.K. Baharuddin. 1986. Relationship of soil dispersibility to infiltration and erosion of southeastern soils. *Soil Sci.* 142:235–240.

Miller, W.P., and J. Scifres. 1988. Effect of sodium nitrate and gypsum on infiltration and erosion of a highly weathered soil. *Soil Sci.* 148:304–309.

Milthorpe, P.L., and J.C. Newman. 1979. Gypsum assists reclamation of scalded sodic clay soils near Condobolin. *J. Soil Conserv. Serv. NSW* 35:149–155.

Morin, J., and Y. Benyamini. 1977. Rainfall infiltration into bare soils. *Water Resour. Res.* 13:813–817.

Morin, J., Y. Benyamini, and A. Michaeli. 1981. The dynamics of soil crusting by rainfall impact and the water movement in the soil profile. *J. Hydrology* 52:321–335.

Morin, J., S. Goldberg, and I. Seginer. 1967. A rainfall simulator with a rotating disk. *Trans. Am. Soc. Agric. Eng.* 10:74–79.

Mortvedt, J.J., D.A. Mays, and G. Osborn. 1981. Uptake by wheat of cadmium and other heavy metal contaminants in phosphate fertilizers. *J. Environ. Qual.* 10:193–197.

Murray, R.C. 1964. Origin and diagenesis of gypsum and anhydrite. *J. Sediment. Petrol.* 34:512–523.

Nakayama, F.S. 1971. Calcium complexing and the enhanced solubility of gypsum in concentrated sodium-salt solutions. *Soil Sci. Soc. Am. Proc.* 35:881–883.

Noble, A.D., M.E. Sumner and M.V. Fey. 1988a. Calcium aluminum balance and the growth of soybean roots in nutrient solution. Soil Sci. Soc. Amer. J. 52 (in press).

Noble, A.D., M.E. Sumner and A.K. Alva. 1988b. Effect of pH on calcium sulfate alleviation of aluminum phytotoxicity. *Soil Sci. Soc. Am. J.* 52 (in press).

Nordstrom, S.K. 1982. The effect of sulfate on Al concentrations in natural waters: some stability relations in the system Al_2O_5-SO_3-H_2O at 298°K. *Cosmochim.* Acta 46:681–692.

Norrish, K. 1954. The swelling of montmorillonite. *Disc. Faraday Soc.* 18:120–134.

Northcote, K.H., and J.K.M. Skene. 1972. Australian soils with saline and sodic properties. *CSIRO Aust. Div. Soils. Publ. 27.*

Oates, K.M., and A.G. Caldwell. 1985. Use of by-product gypsum to alleviate soil acidity. *Soil Sci. Soc. Am. J.* 49:915–918.

O'Brien, L.O., and M.E. Sumner. 1988. Effect of phosphogypsum on leachate and soil chemical composition. *Commun. Soil Sci. Plant Anal.* 19 (in press).

Oliveira, I.P., J. Kluthcouski, and F.N. Reynier. 1986. Efeito de Fosfogesso na Produção de Feijão e Arroz e no Comportamento de Alguns Nutrientes. pp. 45–60. In: *An. Sem. Uso Fosfogesso Agricultura*, EMBRAPA, Brasilia D.F., Brazil.

Onofiok, O., and M.J. Singer. 1984. Scanning electron microscopy studies of surface crusts formed by simulated rainfall. *Soil Sci. Soc. Am. J.* 48:1137–1143.

Orvedal, A.C., and K.T. Ackerson. 1972. *Agricultural Soil Resources of the World.* U.S. Dept. of Agriculture, Soil Conservation Service, Washington, D.C. (mimeographed).

Oster, J.D. 1982. Gypsum usage in irrigated agriculture: A review. *Fertilizer Research* 3:73–89.

Oster, J.D., and H. Frenkel. 1980. The chemistry of the reclamation of sodic soils with gypsum and lime. *Soil Sci. Soc. Am. J.* 44:41–45.

Oster, J.D., and F.W. Schroer. 1979. Infiltration as influenced by irrigation water quality. *Soil Sci. Soc. Am. J.* 43:444–447.

Oster, J.D., and M.J. Singer. 1984. *Water Penetration Problems in California Soils.* University of California, Davis, California.

Oster, J.D., I. Shainberg, and J.D. Wood. 1980. Flocculation value and gel structure of Na/Ca montmorillonite and illite suspension. *Soil Sci. Soc. Am. J.* 44:955–959.

Palmer, B., M. McCaskill, D. Friesen, and L. Hammond. 1983. Sulfur containing fertilizers, past, present, and future. In: Blain, G.J., and A.R. Till (Eds.), *Sulfur in S.E. Asian and S. Pacific Agriculture*, pp. 301–314. UNE, Indonesia.

Palmer, R.S. 1964. The influence of a thin water layer on water drop impact forces. *Int. Assoc. Sci. Hydrol. Publ.* 65:141–148.

Pavan, M.A., and F.T. Bingham. 1986. Effects of phosphogypsum and lime on yield, root density and fruit and foliar composition of apple in Brazilian Oxisols. *Proc. 2nd Int. Symp. Phosphogypsum.* University of Miami, Miami, Florida.

Pavan, M.A., F.T. Bingham, and F.J. Peryea. 1987. Influence of calcium and magnesium salts on acid soil chemistry and calcium nutrition of apple. *Soil Sci. Soc. Am. J.* 52:1526–1530.

Pavan, M.A., F.T. Bingham, and P.F. Pratt. 1982. Toxicity of Al to coffee in Ultisols and Oxisols amended with $CaCO_3$, $MgCO_3$ and $CaSO_4 \cdot 2H_2O$. *Soil Sci. Soc. Am. J.* 46:1201–1207.

Pavan, M.A., F.T. Bingham, and P.F. Pratt. 1984. Redistribution of exchangeable Ca, Mg and Al following lime or gypsum applications to a Brazilian Oxisol. *Soil Sci. Soc. Am. J.* 48:33–38.

Pearson, R.W., F. Abruna, and J. Vincente-Chandler. 1962. Effect of lime and N applications on downward movement of Ca and Mg in two humid tropical soils of Puerto Rico. *Soil Sci.* 93:77–82.

Pearson, R.W., J. Childs, and Z.F. Lund. 1973. Uniformity of limestone mixing in acid subsoils as a factor in cotton root penetration. *Soil Sci. Soc. Am. Proc.* 37:727–732.

Peele, T.C. 1936. The effect of calcium on the erodibility of soils. *Soil Sci. Soc. Am. Proc.* 1:47–58.

Perez-Escolar, R., and M.A. Lugo-Lopez. 1978. Effect of depth of lime application on yield of two corn hybrids grown on a typical Ultisol of Puerto Rico. *J. Agric. Univ. Puerto Rico* 62:203–213.

Poesen, J. 1987. The role of slope angle in surface seal formation. In: V. Gardiner (ed.), *International Geography*, Part II. Wiley, New York.

Ponnamperuma, F.N. 1982. Genotypic adaptability as a substitute for amendments on toxic and nutrient deficient soils. *Proc. 9th Int. Plant Nut. Coll.* 2:467–473.

Quaggio, J.A., A.R. Dechen, and B. van Raij. 1982. Efeitos da Applicação de Calcario e Gesso Solve a Produção de Amendoim e Lixiviacao de Bases no Solo. *R. Bras. Ci. Solo* 6:189–194.

Quirk, J.P., and R.K. Schofield. 1955. The effect of electrolyte concentration on soil permeability. *J. Soil Sci.* 6:163–178.

Radcliffe, D.E., R.L. Clark, and M.E. Sumner. 1986. Effect of gypsum and deep rooting perennials on subsoil mechanical impedance. *Soil Sci. Soc. Am. J.* 50:1566–1570.

Rechcigl, J.E., R.B. Reneau, and D.E. Starner. 1985. Effect of subsurface amendments and irrigation on alfalfa growth. *Agron. J.* 77:72–75.

Reeve, N.G., and M.E. Sumner. 1972. Amelioration of subsoil acidity in Natal Oxisols by leaching surface applied amendments. *Agrochemophysica* 4:1–6.

Rengasamy, P., R.S.B. Greene, G.W. Ford, and A.H. Mehanni. 1984. Identification of dispersive behavior and the management of red brown earths. *Aust. J. Soil Res.* 22:413–432.

Rhoades, J.D., D.B. Krueger, and M.J. Reed. 1968. The effect of soil mineral weathering on the sodium hazard of irrigation waters. *Soil Sci. Soc. Am. Proc.* 32:643–647.

Ritchey, K.D., J.E. Silva, and U.F. Costa. 1982. Calcium deficiency in clayey B horizons of Savanna Oxisols. *Soil Sci.* 133:378–382.

Ritchey, K.D., D.M.G. Souza, E. Lobato, and O. Correa. 1980. Calcium leaching to increase rooting depth in a Brazilian Savannah Oxisol. *Agron. J.* 72:40–44.

Rose, C.W. 1985. Developments in soil erosion and deposition models. In: *Advances in Soil Science*, Vol. 2, pp. 1–63. Springer-Verlag, New York.

Rosolem, C.A., and J.R. Machado. 1984. Efeitos de Calagem e Gessagem no Produção de Algodão e na Lixiviação de Bases em Dois Latossolos. *R. Bras. Ci. Solo.* 8:103–109.

Roth, C.H., and M.A. Pavan. 1988. The effect of lime and gypsum on clay dispersion, infiltrability and chemical composition of runoff and leachate of a Typic Haplorthox. *Geoderma* (in press).

Roth, C.H., M.A. Pavan, and J.C.D. Chaves. 1984. Efcito da Applicao de Calcario e Gesso em Cafceiros Sobre a Infiltracao e Esorrimento em Latossolo Roxo. *Proc. XI Congin. Bras. Pesqu.*, Cafeciros, Lonbina, Brazil.

Russo, D., and E. Bresler. 1981. Soil hydraulic properties as stochastic processes: I. An analysis of field spatial variability. *Soil Sci. Soc. Am. J.* 45:682–697.

Sanchez, P.A. 1977. Advances in the management of Oxisols and ultisols in tropical South America. In: *Proceedings, International Seminar on Soil Environment and Fertility Management in Intensive Agriculture*, pp. 535–566. Society of the Science of Soil and Manure, Tokyo, Japan.

Sanchez, P.A., and J.G. Salinas. 1981. Low input technology for managing Oxisols and Ultisols in Tropical America. *Adv. Agron.* 34:280–398.

Schofield, R.K., and H.R. Samson. 1954. Flocculation of kaolinite due to the attraction of oppositely charged crystal faces. *Faraday Dis. Chem. Soc.* 18:135–145.

Shainberg I., I. Bresler, and Y. Klausner. 1971. Studies on Na/Ca montmorillonite systems. I. The swelling pressure. *Soil Sci.* 111:214–219.

Shainberg, I., R. Keren, and H. Frenkel. 1982. Response of sodic soils to gypsum and $CaCl_2$ application. *Soil Sci. Soc. Am. J.* 46:113–117.

Shainberg, I., and J. Letey. 1984. Response of soils to sodic and saline conditions. *Hilgardia* 52:1–57.

Shainberg, I., and H. Otoh. 1968. Size and shape of montmorillonite particles saturated with Na/Ca ions. *Isr. J. Chem.* 6:251–259.

Shainberg, I., and M.J. Singer. 1985. Effect of electrolyte concentration on the hydraulic properties of depositional crust. *Soil Sci. Soc. Am. J.* 49:1260–1263.

Shainberg, I., J.D. Rhoades, and R.J. Prather. 1981a. Effect of low electrolyte concentration on clay dispersion and hydraulic conductivity of a sodic soil. *Soil Sci. Soc. Am. J.* 45:273–277.

Shainberg, I., J.D. Rhoades, D.L. Suarez, and R.J. Prather. 1981b. Effect of mineral weathering on clay dispersion and hydraulic conductivity of sodic soils. *Soil Sci. Soc. Am. J.* 45:287–291.

Shalhevet, J., and J. Kamburov. 1976. Irrigation and salinity worldwide survey. *Int. Commun. Irrig. Drainage*, New Delhi, India.

Shanmuganathan, R.T., and J.M. Oades. 1983a. Influence of anions on dispersion and physical properties of the A horizon of a red brown earth. *Geoderma* 29:257–277.

Shanmuganathan, R.T., and J.M. Oades. 1983b. Modification of soil physical properties by addition of calcium compounds. *Aust. J. Soil Res.* 21:285–300.

Sherard, J.L., L.P. Dannigan, R.S. Decker, and E.F. Steele. 1976. Pinhole test for identifying dispersive soils. *J. Geotech. Eng. Div. ASCE* 102:69–85.

Shomer, I., and U. Mingelgrin. 1978. A direct procedure for determining number of plates in tactoids of smectites: The Na/Ca montmorillonite case *Clays Clay Miner.* 26:135–138.

Simpson, J.R., A. Pinkerton, and J. Lazdovskis. 1979. Interaction effects of subsoil acidity and water on the root behavior and root growth of some genotypes of lucerne. *Aust. J. Agric. Res.* 30:609–619.

Sinclair, A.G., and W.M.H. Saunders. 1982. Sulphur fertilizers in New Zealand. *Sulphur Agric.* 6:15–22.

Singer, M.J., P. Janitzky, and J. Blackard. 1982. The influence of exchangeable sodium percentage on soil erodibility. *Soil Sci. Soc. Am. J.* 46:117–121.

Smal, H., M.E. Sumner, A. Csinos and C. Kvien. 1988. On the Ca nutrition of peanut (*Arachis hypogaea* L.): II. Diagnostic methods. Unpublished manuscript. University of Georgia, Athens, Georgia.

Smith, D.D., and W.H. Wischmeier. 1957. Factors affecting sheet and rill erosion. *Trans. Am. Geophys. Union* 38:889–896.

Southard R.J., I. Shainberg, and M.J. Singer. 1988. Influence of electrolyte concentration on the micromorphology of artificial depositional crust. *Soil Sci.* (in press).

Souza, D.M.G., and K.D. Ritchey. 1986. Uso do gesso no solo de cerrado. pp. 119–144. *In: An. Sem. Uso Fosfogesso Agricultura*, EMBRAPA, Brasilia D.F., Brazil.

Sposito, G. 1985. Chemical models of weathering in soils. In: J.I. Drever (Ed.), *The Chemistry of Weathering*. D. Reidel, New York.

Sposito, G. and S.V. Mattigod. 1979. Geochem: A computer program for the calculation of chemical equilibria in soil solutions and other natural water systems. Kearney Foundation of Soil Science, University of California, Riverside, CA.

Stul, M.S., and W.J. Mortier. 1974. The heterogeneity of the charge density in montmorillonites. *Clays Clay Miner.* 22:391–296.

Sumner, M.E. 1970. Aluminum toxicity—A growth limiting factor in some Natal sands. *Proc. S. Afr. Sugar. Technol. Assoc.* 44:197–203.

Sumner, M.E., and E. Carter. 1988. Amelioration of subsoil acidity. *Commun. Soil Sci. Plant Anal.* 19 (in press).

Sumner, M.E., M.V. Fey, and M.P.W. Farina. 1987. Amelioration of acid subsoils with phosphogypsum. *Proc. 2nd Int. Symp. Phosphogypsum*, pp. 41–45, University of Miami, Miami, Florida.

Sumner, M.E., W.P. Miller, D.E. Radcliffe, and M. McCray. 1985. Use of phosphogypsum as an amendment for highly weathered soils. *Proc. 3rd Int. Phosphogypsum Symp.*, pp. 111–136. Florida Institute for Phosphate Research, Bartow, Florida.

Sumner, M.E., H. Shahandeh, J. Bouton, and J. Hammel, 1986. Amelioration of an acid soil profile through deep liming and surface application of gypsum. *Soil Sci. Soc. Am. J.* 50:1254–1258.

Syed, O.S.R. 1987. Deleterious effects of gypsum amendments on potassium and magnesium status of field soils. M.S. Thesis, University of Georgia, Athens, Georgia.

Tanaka, A., T. Tadano, K. Yamamoto, and N. Kanamura. 1987. Comparison of toxicity to plants among Al^{3+}, $AlSO_4^+$, and Al-F complex ions. *Soil Sci. Plant Nutr.* 33:43–56.

Tanji, K.K., and S.J. Deverel. 1984. Simulation modeling for reclamation of sodic soils. In: I. Shainberg and J. Shalhevet (eds.), *Soil Salinity under Irrigation*, pp. 238–251. Springer-Verlag, New York.

Tanji, K.K., L.D. Doneen, G.V. Ferry, and R.S. Ayers. 1972. Computer simulation analysis on reclamation of salt-affected soil in San Joaquin Valley, California. *Soil Sci. Soc. Am. Proc.* 36:127–133.

Trochoulias, T., P. Broadbent, and D.R. Baigent. 1986. Response of avocado to calcareous and organic amendments. *Acta Hortic. (The Hague)* 175:179–181.

U.S. Salinity Laboratory Staff. 1954. Diagnosis and Improvement of Saline and Alkali Soils. *USDA Agric. Handb. No. 60.*

van der Boon, J. 1980. Prediction and control of bitter pit in apples. II. Control by summer pruning, fruit thinning, delayed harvesting, and soil calcium dressing. *J. Hortic. Sci.* 55:313–321.

Van Olphen, H. 1977. *An Introduction to Clay Colloid Chemistry* 2d Ed. Wiley, New York.

Velasco-Molina, H.A., A.R. Swoboda, and C.L. Godfrey. 1971. Dispersion of soils of different mineralogy in relation to sodium adsorption ratio and electrolyte concentration. *Soil Sci.* 111:282–287.

Vieira, S.R., D.R. Nielsen, and J.W. Biggar. 1981. Spatial variability of field-measured infiltration rate. *Soil Sci. Soc. Am. J.* 45:1040–1048.

Vilela, H. 1986. Gesso e fosfato natural na recuperação de pastagem. In: *Encontro Tecnico sobre Gesso Agricola*. Belo Horizonte, Brazil.

Vitti, G.C., M.E. Ferreira, and E. Malavolta. 1986. Respostas de culturas anuais e perenes. pp. 17–43. In: *I Seminário sobre o Uso do Fosfogesso na Agricultura.* EMBRAPA, Brasilia, Brazil.

Walker, P.H., J. Hutka, A.J. Moss, and P.I.A. Kinnell. 1977. Use of a versatile experimental system for soil erosion studies. *Soil Sci. Soc. Am. J.* 41:610–612.

Walker, M.E., R.A. Flowers, R.J. Henning, T.C. Kresling, and B.G. Mullinix. 1979. Response of early bunch peanuts to calcium and potassium fertilization. *Peanut Sci.* 6:119–123.

Walworth, J.L., and M.E. Sumner. 1987. The Diagnosis and Recommendation Integrated System (DRIS). *Advances in Soil Science*, Vol. 6, pp. 149–188. Springer-Verlag. New York.

Wang, C.H. 1978. Sulphur fertilization of rice. *Sulphur Agric.* 2:13–16.

Warrick, A.W., and D.R. Nielsen. 1980. Spatial variability of soil physical properties in the field. In: D. Hillel (ed.), *Applications of Soil Physics*, pp. 319–344. Academic Press, New York.

Warrington, D., I. Shainberg, M. Agassi, and Y. Morin. 1988. Effect of slope and phosphogypsum on runoff and erosion. *Soil Sci. Soc. Am. J.* 52 (in press).

Wischmeier, W.H., and J.M. Mannering. 1969. Relation of soil properties to its erodibility. *Soil Sci. Soc. Am. Proc.* 33:131–147.

Wischmeier, W.H., and D.D. Smith. 1978. Predicting rainfall erosion losses. *Agric. Handbk. No. 537*, U.S. Dept. of Agriculture, Washington, D.C.

Wolf, J.M. 1975. Soil-water relations in Oxisols of Puerto Rico and Brazil. In: E. Bornemisra, and A. Alvardo (eds.), *Soil Management in Tropical America*, pp. 145–154. North Carolina State University, Raleigh, North Carolina.

Yaron, B., and G.W. Thomas. 1968. Soil hydraulic conductivity as affected by sodic water. *Water Resour. Res.* 4:545–552.

Zhang, G.Y., X.N. Zhang, and T.R. Yu. 1987. Adsorption of sulfate and fluoride by variable charge soils. *J. Soil Sci.* 38:29–38.

Heavy Metals in Soils and Their Environmental Significance

K.G. Tiller

I. Introduction 113
A. General 113
B. Terminology 115
II. Sources of Pollution 116
A. Deliberate Additions of Heavy Metals to Agricultural Soils 117
B. Inadvertent Transfers to the Soil–Plant–Animal System 121
III. Mobility of Metals in Soils and Landscapes 126
IV. Ecological Consequences of Heavy Metal Pollution of Soils 128
A. Effects on Biological Activity 128
B. Effect on Soil–Crop–Animal System 128
V. Prediction of the Impact of Heavy Metal Pollution of Soils 130
A. Degree of Soil Pollution 130
B. Prediction of Uptake of Metals by Plants by Means of Soil Analysis 130
C. Prediction by Plant Analysis 131
D. Measurement of Regional Variations of Heavy Metal Status of Soils 132
E. Background Levels of Heavy Metal Pollution 132
VI. Guidelines for Regulatory Control of Pollution of Agricultural Land .. 133
A. Experimental Approaches 133
B. Legislative Approaches 136
VII. Final Comment 137
References 137

I. Introduction

A. General

Modern civilization is completely dependent on a large range of metals for all aspects of daily life. There is a long history of association between metals and human development (Encyclopedia Britannica, 1980). Copper has been used

Table 1. Worldwide natural and anthropogenic emissions

		Heavy metal emissions (× 10^3 tonnes)				
Emission	Period	Cd	Cu	Ni	Pb	Zn
Natural	Annual	0.83	18	26	24	44
Anthropogenic	Annual	7.3	56	47	450	310
Anthropogenic	Total	316	2,180	1,000	19,600	14,000

Source: Nriagu (1979).

since about 8000 B.C., initially as native copper, but smelting of copper from oxide ores is thought to date from about 6000 B.C. Lead was used before 5000 B.C., zinc and mercury by about 500 B.C., and nickel in alloys by 200 B.C. Cadmium was discovered comparatively recently (1817). The chronology of the environmental impact of heavy metals has been assessed by examination of cores from sediments (Müller, 1981), ice (Murozumi et al., 1969), and peat (Chamberlain, 1983). A peak in lead usage was noted at the time of the Roman Empire; metal use accelerated during the Industrial Revolution of the nineteenth century, and since then heavy metals have become 'essential' to modern society because of the range of metal products used. All-time usage of Cd, Cu, Pb, Ni, and Zn has been estimated to have been 0.5, 307, 241, 17, and 250 × 10^9 kg, respectively (Nriagu, 1979). The estimated total (all-time) and annual amounts of these metals dispersed through the atmosphere are presented in Table 1. This enormous input into the atmosphere is increased by liquid effluents and direct deposition of solid wastes onto land and into water.

Our increasing world population and the increasing annual metal usage per capita leads inevitably to ecological problems because of wide dispersal of metals into the natural environment. The heavy metals, which occur naturally in all soils, waters, and living organisms, can have severe toxic impact on living systems in excessive concentrations, but in fact many are essential for healthy life. It is critically important to recognize the boundaries of concentration that separate healthy from unhealthy organisms or safe food products from those unsuited for animal or human consumption. We also need to recognize the variations of chemical reactivity and mobility of each pollutant metal in different natural environments because they affect redistribution in soils, landscapes, the hydrological system, and passage through the food chain. These differences need to be recognized so that oversimplification can be avoided when assessing the ecological consequences of any polluting activity and providing guidelines for safe agricultural practice. Dispersal of heavy metals and their largely irreversible retention by soils have long-term consequences. Pollution sources clearly need to be minimized, but also reactions of metals with soils must be thoroughly understood to accurately assess the future problems of food production and quality and to plan necessary countermeasures.

Heavy metal pollution is emitted from various sources as solids, liquids, and, frequently, as fine particles directly into the atmosphere, but is ultimately deposited on the surface of water bodies and land. In the resultant dispersal pattern, locally more polluted areas are superimposed on a broad regional or even continental smear of contamination. These geographic variations need to be recognized.

Pollution on a local or limited regional scale, arising from identifiable sources, can arouse community concern and has often stimulated research and remedial action. The possible effect of the continuing, irreversible transfer of massive amounts of potentially toxic metals from localized ore bodies, coal seams, etc. to the earth's surface on a global scale is also cause for concern. Unfortunately the stealth of advancing contamination, and its lack of respect for national borders, weakens the resolve for coordinated countermeasures.

In this treatise, I review some general aspects of heavy metal pollution, with emphasis on Cd, Pb, and Zn. Some of the main features of heavy metal pollution of soils and the consequences are highlighted. Examples and case histories will be biased by my research experience in Australia. Soil pollution will be viewed as a process that adversely affects soil productivity in terms of quantity and quality of the products.

B. Terminology

Discussions of heavy metals in soils and plants are sometimes complicated by terminology. Many of the "essential" and "toxic" elements are ubiquitous in nature and previously they could be measured only with difficulty and often barely detected. Consequently, elements were termed trace elements. The term continues in use even though these elements can now be analyzed quickly and accurately in much lower concentrations than were possible several decades ago. Trace elements, while still a useful description of those elements present in soils and plants in small amounts, does not imply their physiological role in living things. This term has other drawbacks; for example, an element occurring at a low concentration in the bulk soil, say 10–100 mg kg^{-1}, may have a very high concentration on a microscale, say $>10\%$. For those elements known to be essential in animal, human, and/or plant nutrition, "micronutrients" is a more useful and explicit term. Some elements may stimulate growth of an organism although evidence of essentiality is lacking. The physiological function of some elements depends on concentration; that is, they are essential for life or are passively absorbed without physiological function at low concentrations, but have a deleterious effect on health at higher contractions, at which they may be termed toxic.

The micronutrients of importance for plant growth are $\underline{Cu}$, $\underline{Zn}$, $\underline{Fe}$, $\underline{Mn}$, $\underline{Mo}$, $\underline{Co}$, Na, B, Cl, and possibly $\underline{V}$ (Nicholas, 1975). Not all elements are essential to all plants For example, only C_4 plants respond to Na (Brownell and Crossland, 1972). There is increasing evidence that Ni may be essential for the growth and reproduction of plants (Hutchinson, 1981; Walker et al., 1985) and that some rare earth metals may stimulate crop growth (Zhu and Liu, 1986). For animal health, the following trace elements are essential: Cu, Zn, Fe, Mn, Co, Mo, Cr, Sn, Si, Ni, I, Se, F, V, and As (Underwood, 1977). Many of these elements (those underlined) essential for animal and plant life are considered "heavy metals."

In the context of environmental studies about metals. the terms "toxic" and "heavy", or both combined, have been commonly applied. These terms are less satisfactory than trace or micronutrient, as just discussed, because nearly all substances, including water, are toxic at sufficiently high concentrations, and heavy has no accepted basis of definition, whether in reference to specific gravity or atomic weight. By common usage, heavy metals appear to include all metals of the periodic table with atomic numbers greater than 20, generally excluding the alkali metals and the alkali earths. (The heavy metals so defined constitute two thirds of the periodic table.) However, this imprecise term is commonly used in the literature and will probably continue to provide a useful umbrella term for metals classed as environmental pollutants. The heavy metals considered most important as environmental pollutants are Cd, Pb, Cu, Zn, Ni, Sb, and Bi. The term heavy metal is least satisfactory for those metals involved in biochemical and toxicological processes. Categories connected with ionic properties, especially for complex formation, are then preferable. Phipps (1981) has discussed metal classes based on Lewis acid properties.

II. Sources of Pollution

The main factors determining the metal content of the soil are usually the composition of the parent material and the pedogenic processes in soil formation. Many studies have shown that even uniform parent materials can result in soils with quite different trace metal concentrations. For example, Tiller (1963) showed that the total Ni concentration in surface soils on dolerite in Tasmania ranged from 5 to 140 mg kg^{-1}. The diverse and complex contributions of heavy metals from anthropogenic sources are superimposed on the variations in natural or background levels which needs to be recognized in the formulation of guidelines relating to maximum permissible levels of metals in soils. These guidelines are discussed later.

Contamination of soils by metal-containing substances can result from either the decisions of local farm management to use agricultural sprays, fertilizers or waste products as soil amendments, that is, deliberate (farm-initiated) pollution, or from sources incidental to modern society, such as factory and power station emission, which are often beyond local control (inadvertent pollution). These anthropogenic sources of pollution (Figure 1) should be considered in context with natural sources such as windblown dusts, volcanoes, forest fires, etc. Nriagu (1979) has estimated (Table 1) that anthropogenic emissions of heavy metals into the atmosphere appreciably exceed natural emissions of Cu, Ni, Zn, Cd, and Pb and by an order of magnitude for Cd and Pb.

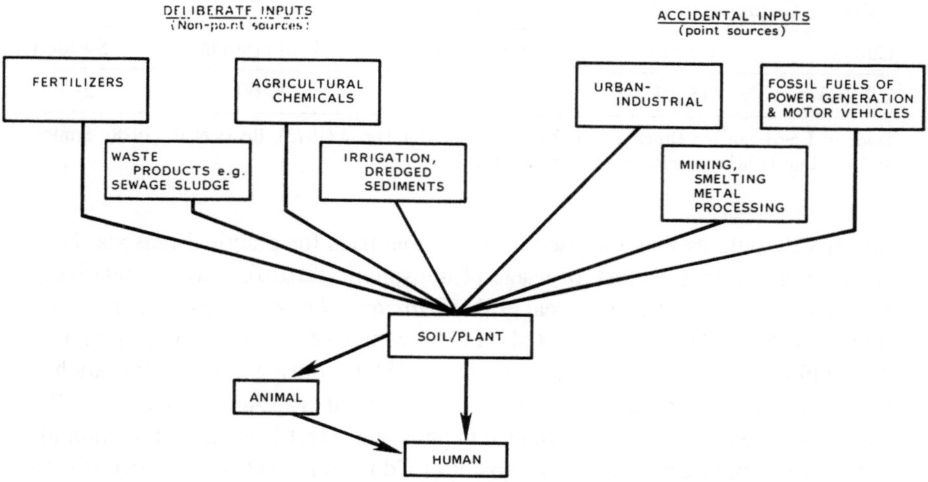

Figure 1. Anthropogenic pollution of the soil–plant–animal system.

A. Deliberate Additions of Heavy Metals to Agricultural Soils

Many additions of heavy metals to agricultural land are farmer initiated. Practical alternatives are often not possible because economic production depends on use of the available agricultural chemicals and fertilizers, and irrigation waters.

1. Agricultural Chemicals and Fertilizers

a. Fertilizers

Although micronutrient heavy metals are often added to fertilizers as needed, the major nutrient fertilizers often contain appreciable concentrations as incidental impurities derived from the raw materials, which are sometimes sufficient to benefit crop growth (Tiller, 1983). Commercial fertilizers may contain a great range of heavy metals (Swaine, 1962), but the highest values for heavy metals are found in phosphatic fertilizers. Ranges in concentrations reported in several countries are shown in Table 2. These values reflect the worldwide range of Cd concentrations in rock phosphates of 1–90 mg kg^{-1} (Williams and David, 1973; Baechle and Wolstein, 1984).

In recent decades, particular concern has been focussed on Cd in fertilizers because of the relative ease of transfer of Cd from soil to plant and the perceived health risk (Schroeder and Balassa, 1963; Williams and David, 1973). Entry of fertilizer-derived Cd into the food chain depends on the amount of fertilizer used, its Cd content, fertilizer and soil properties (especially pH), and the crop's ability to absorb Cd (Smilde and van Luit, 1983). Of these factors, the Cd content as determined by the rock phosphate source is the most difficult to control, because supplies of rock phosphate with low Cd contents are finite and sources with

Table 2. Cadmium in some phosphatic fertilizers (in mg kg^{-1})

Canada	Australia	United States	Netherlands	Sweden
2.1–9.3	18–91	7.4–156	9–60	2–30

Sources: Freedman and Hutchinson (1981); Williams and David (1973); Reuss et al. (1978); Smilde and van Luit (1983); Stenström and Vahter (1974).

higher Cd contents must be used in many countries for practical reasons. Notwithstanding a long history of usage of phosphatic fertilizers with a relatively high Cd content, Australian wheat grain, *Triticum vulgare*, ranges only from less than 5 to 45 μg Cd kg^{-1} (Tiller et al., 1975). Mortvedt (1987) reported that normal applications of P fertilizers in the United States during at least 50 years has not increased Cd concentrations in their main cereal crops and in soybean, *Glycine max*. The current hazard from Cd in phosphatic fertilizers appears minimal, but requires monitoring. The net transfer of Cd from phosphate rock deposits to agroecosystems is considerable; it is disseminated much more widely than other potentially contaminated soil amendments, such as sewage sludge, with which its impact is often compared. The impact of the spread of Cd in fertilizers is more akin to the dispersal of metals by aerial transport and deposition, and in some countries both atmospheric emissions and fertilizers make similar contributions to soil Cd (Gunnarsson, 1983).

b. Agricultural Chemicals

Metal-containing sprays have been regularly used to control pests in horticultural and vegetable crops and in turf. Zinc salts, copper and lead arsenates, and metallo-organic compounds have all been used, and high levels of metal contamination have been documented. In several countries, copper levels of about 1,000 mg kg^{-1} have resulted from extensive use of Bordeaux sprays, that is, copper basic salts (Tiller and Merry, 1982) while use of lead arsenate sprays has led to soil levels of at least 500 mg Pb kg^{-1} and 100mg As kg^{-1} (Freedman and Hutchinson, 1981; Merry et al., 1983). By comparison, uncontaminated soils contained about 20, 10, and 5 mg kg^{-1} for Cu, Pb, and As, respectively (Vinogradov, 1959). Annual accumulations of Cu, Pb, and As in Australian orchard soils were estimated (Merry et al., 1983) at about 4, 5, and 1 mg kg^{-1} respectively, that is, about three orders of magnitude higher than the accumulation rates of Cd from phosphatic fertilizers under European and Australian conditions (2–4 g $ha^{-1}\cdot yr^{-1}$). The main hazard from the contamination of horticultural soils is not usually from existing crops but rather in relation to new uses of the soil after orchard or vineyard clearance.

2. Metal-Contaminated Waste Products

Many urban and farm wastes, when spread on land as a convenient and economic means of dispersal and for the benefit of soil productivity, can cause appreciable contamination by heavy metals.

Table 3. Heavy metal concentrations in sewage sludges

Country	Range and median (mg kg^{-1} dry weight)				
	Cd	Zn	Cu	Pb	Ni
U.K.	2–1500	600–20,000	200–8,000	50–3,600	20–5,300
	(20)	(1,500)	(650)	(400)	(100)
U.S.A.	2–1100	72–16,400	84–10,400	800–26,000	12–2,800
	(12)	(2,200)	(700)	(480)	(52)
Sweden	2–171	700–14,700	52–3,300	52–2,900	15–2,120
	(7)	(1,570)	(560)	(180)	(51)
Canada	2–147	40–19,000	160–3,000	85–4,000	7–1,500
Australia	2–285	240–5,500	250–2,500	55–2,000	20–320
	(26)	(1,900)	(670)	(420)	(60)
C.E.C. Rec. Max	40	3,000	1,500	1,000	400

Note. Median values given in parentheses.

Sources: Davis (1980); Page (1974); Freedman and Hutchinson (1981); de Vries (1983); Webber et al. (1983); C.E.C. (1982).

a. Municipal Wastes

Both composted and uncomposted municipal wastes, which have been added to the soil in several countries, may contain appreciable amounts of heavy metals, as much as 2, 330, 215, and 800 mg kg^{-1} of Cd, Cu, Pb, and Zn, respectively (Purves, 1977). However, this material is not valued as a soil amendment, and any environmental hazard would be related to local use.

b. Pig Waste Slurries

Pig and poultry waste slurries commonly contain 600–900 mg Cu kg^{-1} (dry basis), but can contain as much as 2,000 mg kg^{-1} (Unwin, 1981) and if spread on land on a regular basis could cause pollution problems in some soil and topographic situations. Contamination of pastures with Cu-rich slurries is considered a particular hazard to grazing sheep.

c. Sewage Sludge

Huge amounts of sewage sludge are produced worldwide; for example, 6 million tonnes of sludge (dry basis) are produced annually in the EEC alone (Kofoed, 1984). The disposal of sludge on land has become increasingly favored by city authorities as a means of disposal and welcomed by farmers as a source of nutrients and as a soil conditioner. Unfortunately, sludges contain appreciable amounts of heavy metals that persist in surface soils long after application. Sewage sludges from different parts of the world have similar ranges heavy metal concentration (Table 3).

Metals in sewage sludge most likely to affect crop production are Zn, Ni, Cu, and Cd (Page, 1974). Of these, Cd has raised most concern for human health, and has been documented by numerous investigations. More stringent regulations

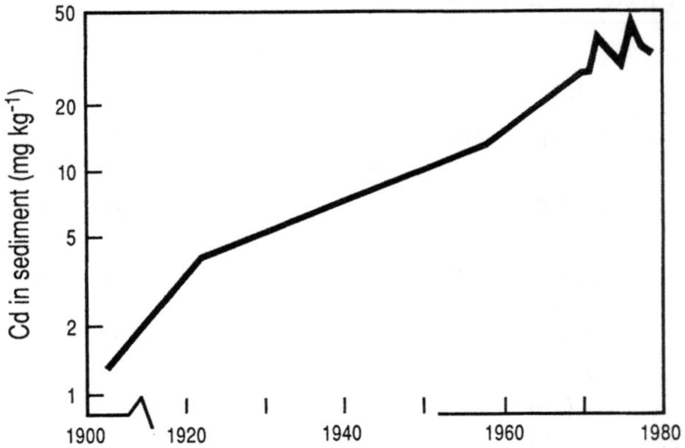

Figure 2. Historical record of Cd pollution of Rhine River sediments (from van Driel et al., 1984).

restricting inputs by industrial effluent have resulted in downward trends in the heavy metal content of sewage sludge, but significant contributions from domestic sources (Klein et al., 1974) will restrict the lower achievable limits. Treatment of soils with sewage sludge invariably results in increased input of heavy metals to the food chain, whether by direct consumption of food crops or indirectly through animal production. The uptake of heavy metals from sludge-amended soils depends on metal concentration and on plant genotype, temperature, dose rate, and equilibration period, and is mediated by soil physicochemical properties, especially pH. These factors have provided a scientific basis for guidelines for the utilization of sludge in many countries, but further research is needed, for example, into the chemical basis of declining availability of metal in sludge-treated soils with time and why this does not always occur. The inability to account for all the Cd added to some soils as sludge, as discussed by Davis (1983), is curious and also deserves attention. Retention of metals in iron oxide minerals (Gerth and Brümmer, 1983) in a form not easily extracted by strong acids, dilution by cultivation and particle migration into the soil profile, and losses of sludge from surface soils by wind action are possible explanations.

3. Miscellaneous Sources

The contamination of some rivers and canals with effluents from mining and industrial activities can also indirectly affect crop quality. Use of contaminated river water for irrigation has led to polluted soils and crops. The occurrence of Itai-itai disease in Japan is one of the best documented examples of this problem (Takijuma and Katsumi, 1973). Dredged sediments from rivers and canals in industrialized areas may have high concentrations of heavy metals. For example, the Rhine River sediments have been increasingly polluted in recent decades and now contain more that 30 mg Cd kg^{-1} (Figure 2) and must be used with caution (van Driel et al., 1984). Periodic flooding of such contaminated rivers and the resultant deposition of polluted sediments would have a similar yet more widespread and less dramatic effect. Heavy metals can be intentional introduced into agrolandscapes from many other minor sources such as metal-containing wood preservatives, coated fencing wire, metal sheeting, etc., but their impact is negligible.

B. Inadvertent Transfers to the Soil–Plant–Animal System

Some of the major inputs of pollution to agricultural land are beyond the control of farm managers and can only be regulated by government authorities. The main sources involved are smelters, metal processing works, power stations, refuse incinerators, and emissions from motor vehicles. These are mainly point sources. Metalliferous and coal-mining activities also pollute land. The dumping of metal-rich overburden, and dispersal of the by-products, for example, tailings, results in very localized soil pollution problems for which a range of renovation procedures has been developed.

1. Urban-Industrial Pollution

Point-source emissions of large metal industries can cause broad-scale contamination of land. The local impact of mining activity of soil may be very high metal concentrations in soils, with values of the order of 1% being reported. Other industries reported to cause significant heavy metal pollution of soils include chemical manufacturing, oil refining, metal processing and plating, tanneries, and fertilizer manufacture. Discussion here is restricted to smelters, power plants, and incinerators.

a. Smelters

Primary metal production is the main source of anthropogenic emission of nonferrous heavy metals, except for lead emissions, which originate principally from motor vehicles (Nriagu, 1979). Dispersal from smelters is strongly dependent on local meteorological and topographical factors in relation to stack height and aerosol particle size. The major point sources exhibit haloes of soil pollution that decrease in concentration with distance. Local soil pollution within 1–2 km may be quite severe, while near-background concentrations of metals in soils may not be recorded at distances of less than about 20 km from the source. Appreciable proportions of emitted metalliferous aerosols escape the smelter vicinity and contribute to global contamination. Numerous studies of pollution around primary metal smelters have been reported (Freedman and Hutchinson, 1981). One of the best documented studies concerns the Sudbury smelters, Ontario, Canada, around which an area of nearly 500 km^2 is severely affected by sulfur dioxide and metal emissions. In this chapter, the impact of heavy metal smelting is illustrated by briefly outlining a case study of a smelter at Port Pirie in South Australia.

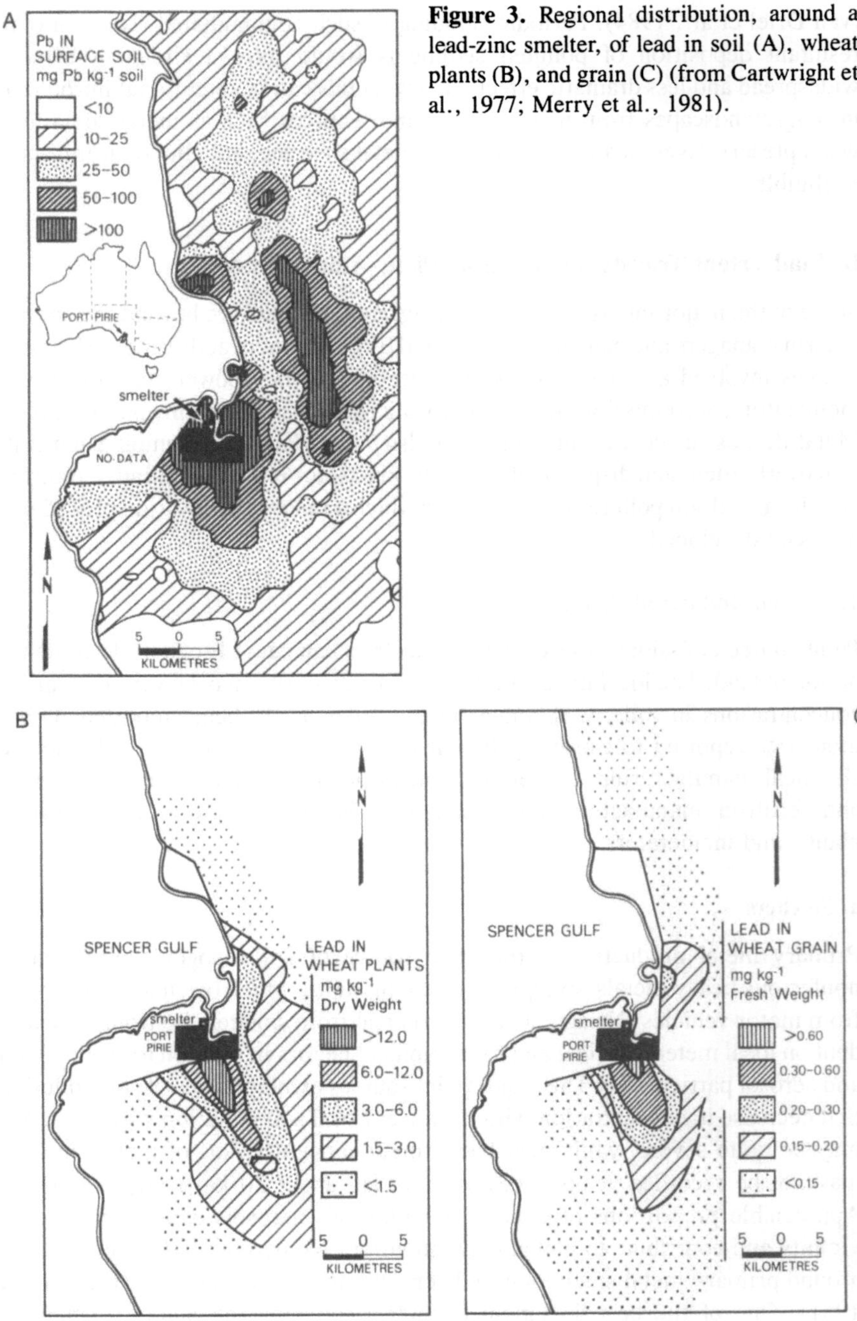

Figure 3. Regional distribution, around a lead-zinc smelter, of lead in soil (A), wheat plants (B), and grain (C) (from Cartwright et al., 1977; Merry et al., 1981).

The Port Pirie lead-zinc smelter, located about 200 km north of Adelaide, was established for the processing of metal sulfide ores from the Broken Hill mines, and since 1889 has produced about 12 million tonnes of lead as well as appreciable amounts of zinc and other metals. The smelter is situated on a coastal plain used for low-intensity mixed cereal/sheep production and vegetable production under irrigation. The study area of 5,000 km^2 extended into a hilly region to the east, which is used for sheep grazing, forestry, and water catchment in State parks. The large smelter is the only point source in an otherwise nonindustrial area of low population density; this facilitated the interpretation of soil contamination and its consequences.

At least 3,400 km^2 were contaminated to some extent (Cartwright et al., 1977). Distributions of Cd, Pb, and Zn in surface soils within the region were very similar; the data for Pb are shown as an example in Figure 3. Contamination declined exponentially with distance on the plain, and was detected as far as 65 km from the smelter depending on direction. The pattern of soil pollution was determined by seasonal wind directions and topography and provided an historic record of metal pollution until the 1940s. Improved smelter design has since drastically reduced stack emissions. The impact of soil pollution was also studied in the field and glasshouse in relation to cereal, pasture, and vegetable production (Tiller et al., 1975; de Vries et al., 1975; Tiller et al., 1976); these studies provided the basis for continuing investigations of animal and human health and of the marine environment. Analyses of wheat plants at the flowering stage and of grain at harvest provided biogeochemical indicators of heavy metals, which mirrored the geographical variation in metal contents of soil (Merry et al., 1981). The results for Pb from these analyses are shown in Figure 3. Fortunately, the cereal-growing soils of this area had alkaline pH values that reduced Cd, Zn, and Pb uptake to less than accepted health standard levels, even on the most polluted soils.

Grazing animals were considered to be at greater risk than humans because of the enhanced metal status of pastures caused by both metal uptake and adhering smelter fume and contaminated soil dust in this seasonally dry region. Marked differences in metal status between pasture plants of different genotypes were noted (Figure 4) because of differences in both absorption by their root systems and in leaf morphology, which affected dust adhesion. Whereas the Pb status of pasture plants was mainly associated with adhering contaminated soil dust, the Cd concentrations primarily resulted from physiological uptake; the zinc status derived from both processes (Merry and Tiller, 1978).

The strong dependence of metal contents of pasture on distance from the smelter is shown in Figure 5 for Pb status of *Atriplex* spp. (saltbush). Lead concentrations in samples closer than 4 km ranged from 78 to 540 mg kg^{-1} dry matter, and are probably responsible for earlier reports of animal health problems, especially with horses. Studies of heavy metal status of sheep grazing at increasing distances from the smelter showed a strong distance dependence of metals ingested and metal concentration in blood, bone, liver, and kidney (Koh and Judson, 1986). In contrast to Pb and Cd, the Cu concentrations in liver were

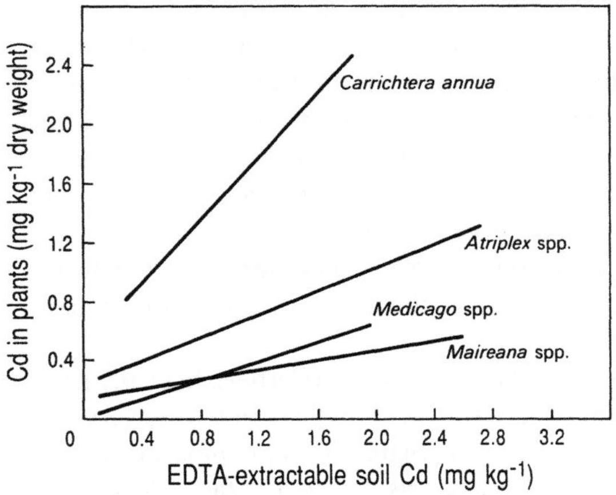

Figure 4. Relation of Cd content of different pasture species to EDTA-extractable soil Cd (from Merry and Tiller, 1978).

inversely related to distance from the smelter (Figure 6) because of interactions with other heavy metals (Suttle, 1975). These results, together with seasonal proliferation of certain pasture plants, explain the occasional occurrences of Cu deficiency in sheep as indicated by wool quality (Hannam and Reuter, 1977).

The distribution of polluted soils and homegrown vegetables in the urban area around the smelter (Tiller and de Vries, 1977) provided the basis of investigations of the incidence of late fetal deaths in this community. Studies by McMichael et al. (1986) showed a significant relationship between maternal blood lead concentrations and the incidence of early birth; investigations of child health and behavior are continuing.

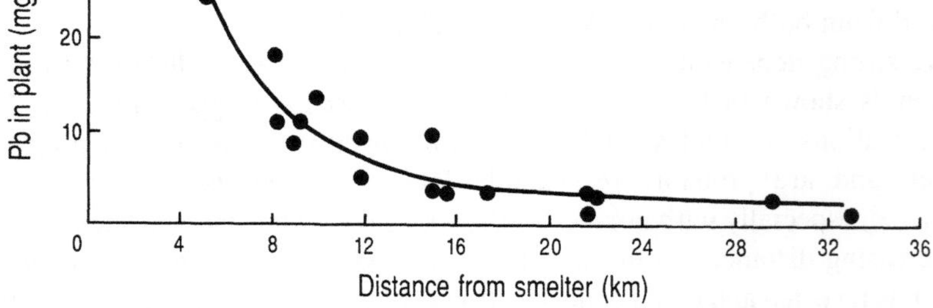

Figure 5. Lead concentration in *Atriplex* spp. (saltbush) related distance from smelter (from Merry and Tiller, 1978).

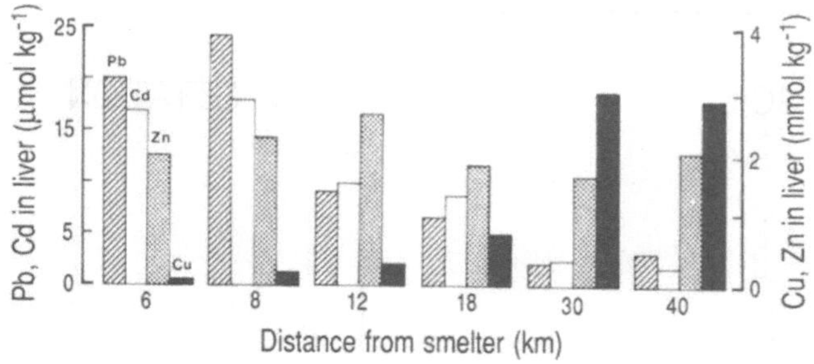

Figure 6. Heavy metal concentrations (Pb, Cd, Cu, Zn) in the liver of sheep related to distance from smelter (from Koh and Judson, 1986).

b. Power Plants and Incinerators

The dependence on coal as a major energy source for electricity generation in many countries results in the production of large quantities of metal-containing particulates (fly ash). Although a major part of fly ash is separated from the exhaust gases, large amounts are still emitted. These fly ashes contain a wide range of metals and other contaminants (Page et al., 1979), but the heavy metal contents are not particularly high in relationship to the normal soil range. Although the smaller particle-sized fractions that tend to have higher metal contents will dispense further, possibly 100 km downwind, Page et al. (1979) consider that the enrichment of soils by stack-derived fly ash to be negligible. Disposal of fly ash on agricultural land is not common. The main hazard from fly ash derives from the more volatile, nonmetallic impurities such as boron.

The incineration of plastics, wood products, fossil fuels, and organic wastes by householders and municipal authorities may also result in atmospheric emission of heavy metals. Låg (1985) has highlighted the possibility of significant pollution of soils within 10 km or so of some municipal incinerators, and urges that the location and design of such facilities be given careful consideration.

2. Emissions from Motor Vehicles

The numerous investigations of heavy metals in the vicinity of busy streets and highways have identified elevated concentrations of heavy metals in soils, plants. soil animals, and air that are related to traffic density, mode of vehicle operation, and distance from the highway. Lead dominates the roadside environment because of its common use in a gasoline additive, lead tetraethyl. Other metals, especially Cd, Zn, Cr, and Ni, also accumulate near roads because of wear of vehicle parts and tires and use of fuel additives. A high proportion of atmospherically emitted Ni is ascribed to nickeliferous diesel oil (Nriagu, 1979). Lead is emitted from vehicle exhausts as fine particles, the coarser particles of which deposit within 50–100 m of the roadway (Figure 7). Most of the emitted Pb particles are <2 μm in diameter and are dispersed much further into the local and

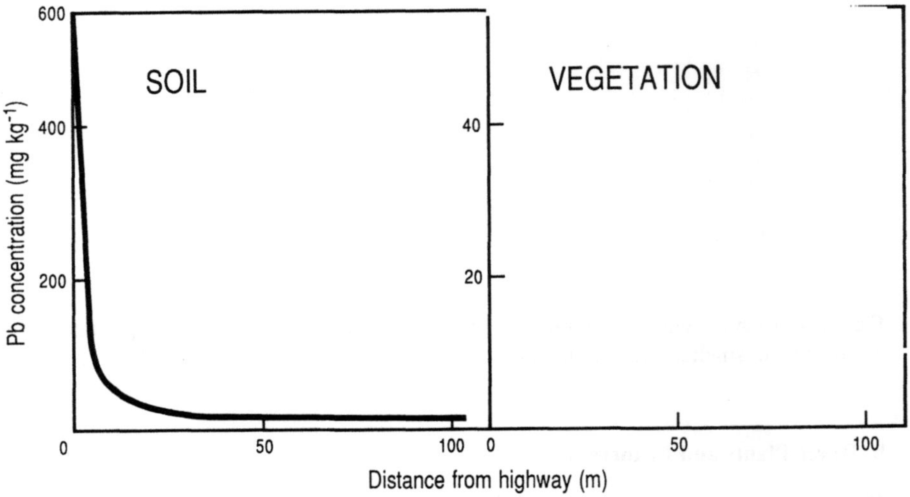

Figure 7. Lead in soil and vegetation related to distance from highway with average traffic of 8,100 vehicles/day (from Wheeler and Rolfe, 1979).

regional environment. Several studies, for example, Page and Ganje (1970), have reported significant contamination by motor vehicles up to several kilometers from the highway. Reiter et al. (1977) concluded that nearly half of automotive Pb emitted was still airborne at a distance of 20 km. Tiller et al. (1987) concluded that at least 30% of automotive lead burned within the city of Adelaide was dispersed from the highway zone and also beyond the study region (>80 km form the city) (Figure 8). The land downwind from the urban source was shown by lead isotope studies (Gulson et al., 1981) to be contaminated by automotive lead to about 50 km, but deposition on the landscape beyond the highway zones only amounted to about 4% of the total lead burned in gasoline (Tiller et al., 1987).

III. Mobility of Metals in Soils and Landscapes

Movements of heavy metals within landscapes and throughout a region will be mainly associated with the solid phase of earth materials, under the agencies of wind, water, and gravity. Long-distance transport of heavy metals occurs largely as fine particles with appreciable residence times in the atmosphere, and is evidenced by the accumulation of heavy metals in the Greenland ice cap (Murozumi et al., 1969) and in the moss carpets of northern Scandinavia (Tyler, 1972). Such transfers demonstrate the slow buildup of soil pollution in areas remote from centers of industrial and urban activity. Within individual catchments, transfer of heavy metals in the solid phase is associated with erosional processes.

Movement of heavy metals within the soil mass will be principally in the solution phase. Hence the chemical factors that control the distribution of metals

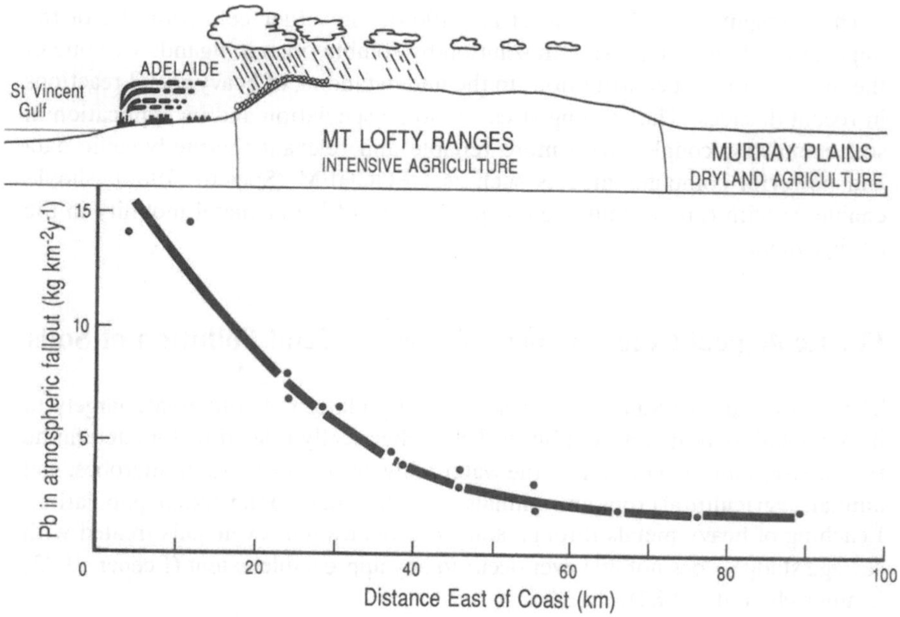

Figure 8. Dispersal of lead from the Adelaide, Australia, urban area into the rural hinterland (from Tiller et al., 1987).

between solid and solution phase, and the chemical forms in each phase, will influence the mobility of heavy metals. In brief, soil chemical reactions controlling mobility of heavy metals and their uptake by plants are, broadly, adsorption/desorption or solubility/precipitation, as determined by the concentration and ionic species in solution and the kind of surface phases present.

Compounds of heavy metals can be demonstrated under extremes of soil conditions, for example, pH, redox, and metal concentration, but the successful application of solubility criteria to normal soil conditions has been limited to date (Tiller, 1983). The main limitations are the tendency for quasi-equilibrium and metastable mineral forms, the low concentrations of heavy metals, which militate against the formation of pure phases, and the difficulty of measuring ionic species in soil solution. The solubility of heavy metal compounds may be important in soils—they certainly provide a framework of thinking—but generally the poorly defined nature of trace metals at soil surfaces and lack of appropriate stability constants will limit progress in solubility studies for some time. The main benefit to date has been to indicate which simple compounds of trace elements, such as zinc, are unlikely to occur in soils.

Adsorption/desorption has provided, by general consensus, a useful means of describing heavy metal reactions in soils, but the boundary between adsorption and solubility is sometimes experimentally and philosophically obscure and is sensitive to small changes in soil components and soil conditions (Brümmer et al., 1983).

The recognition by Hodgson et al. (1965), and other collaborators, of the important role of complexing in solution by soluble organic ligands was one of the most significant contributions to the understanding of heavy metal reactions in recent decades. The ensuing stress on ionic speciation and its application in soil chemistry, coupled with more reliable and relevant thermodynamic data and powerful computer models such as GEOCHEM (Sposito, 1983), should enable us with time to improve our predictions of heavy metal mobility in the environment.

IV. Ecological Consequences of Heavy Metal Pollution of Soils

The ecological consequences of heavy metal pollution of soils relate largely to heavy metal mobility and solubility. These chemically related factors determine transmission through the soil to the water table, availability to soil microbes, soil animals, agricultural crops and animals, and ultimately to the human population. Leaching of heavy metals through soils to groundwater, even soils treated with sewage sludge, does not however occur to any appreciable extent (Leeper, 1972; Emmerich et al., 1982).

A. Effects on Biological Activity in Soil

Pollution of soils by heavy metals may inhibit microbial enzyme activity and reduce the diversity of populations of microorganisms and soil fauna (Tyler, 1981; Küster and Grün, 1984). The toxicity of heavy metals to earthworms depends on the chemical form of the pollutant, the earthworm species, and soil factors. Of Cd, Cu, Ni, Zn, and Pb, Cd appears the most toxic and Zn and Pb the least, with Cu and Ni intermediate in effect (Lee, 1985). Reductions in populations of soil animals by high concentrations of heavy metals have been documented, for example, in orchard soils (Hirst et al., 1961; Mochizuki et al., 1975) but Hughes et al. (1980) suggested that this may only apply at very high concentrations. Smaller populations of soil fauna may affect soil fertility through their role in bioturbation, turnover, and breakdown of organic matter, and soil structure and infiltration. Contaminated soil animals also introduce heavy metals into the food chain of higher animals.

Soil biological processes considered especially sensitive to heavy metals are mineralization of N and P, cellulose degradation, and possible N_2 fixation; these could significantly affect some natural ecosystems. Although the potential is recognized, as discussed previously, there is little evidence to date that soil biological processes are being affected in most polluted soils by heavy metals (Domsch, 1984).

B. Effect on Soil–Crop–Animal System

At moderate levels of heavy metal concentrations above the deficiency range, most natural soils act as a repository or sink for metals without any obvious

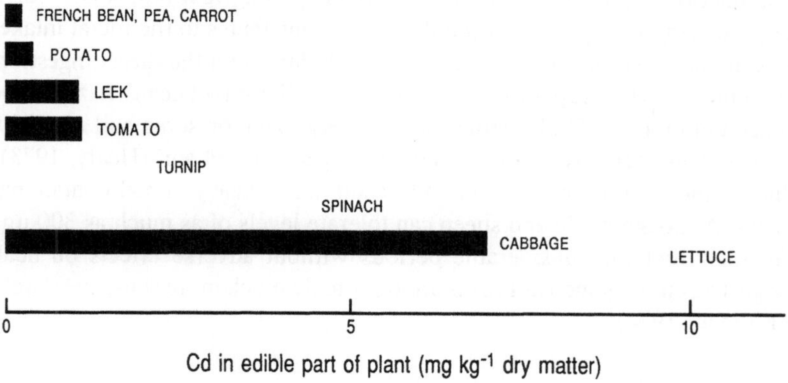

Figure 9. Cadmium concentrations in edible parts of different crops grown in same soil (from Davis and Carlton-Smith, 1980).

effects on soil biological behavior. The important question is the response of plants and animals to increasing concentrations of metals that result from natural variability or from pollution. The extent of transfer of metals from soil to plant depends on the kind of plant (Figure 9), the particular metal involved, and the soil chemical environment.

Uptake of metals by plants reflects increasing levels of metals in soils, especially those metals that are more mobile. Above certain critical concentrations of heavy metals in plants, yield and/or crop quality and animal or human health may be affected. Concentrations of toxic metals require careful monitoring because potentially dangerous levels may be reached in the plant food diet of animals or humans without any evidence of deleterious effects on the plants. This is especially so for Cd, Pb, and Tl, for which plant toxicity occurs above the concentrations considered desirable for human intake. The reverse is true for other heavy metals, for example, Zn, Cu, and Ni, for which the concentrations causing yield depression of plants are below those of risk to humans and animals if the plants are ingested as food. This provides a built-in safety mechanism for such elements.

Heavy metal pollutants rarely occur alone. The associated elements, frequently other heavy metals, and possibly major nutrients also, can affect the biological effectiveness of the metal in question by additive, synergistic, or antagonistic effects. It is clear (see Page et al., 1981) that at least as far as Cd is concerned the interactive effects of different metals cannot easily be generalized. Interactions of Cd with Zn, for example, can be positive, negative or nonexistent, depending on the relative concentration levels, the soil and the crop.

The impact of heavy metal pollution on health also depends on the nature of the association of heavy metals with plants. The heavy metals associated with plants are not necessarily absorbed by the root system and translocated within the plant. In some regions, appreciable amounts (and a high proportion) of some heavy metals can adhere directly to leaf surfaces after deposition from the atmosphere.

Contamination of pasture plants by the spreading of agricultural waste slurries, and by adhering soil splash and soil dust, also contributes to the metal intake by farm animals. Surficial contamination of food plants and the direct ingestion of soil together with forage plants can complicate predictions of health hazard to both animals and humans. Under situations of overgrazing or some soil conditions, animals can incidentally ingest considerable quantities of soil (Healy, 1973).

Unthriftiness in farm animals can result from heavy metal-contaminated pastures. Although cattle and sheep can tolerate levels of as much as 300 μg Pb g^{-1} in their diet for considerable periods without adverse effects on health, monogastric animals such as horses are reportedly much more sensitive (Graham and Kalman, 1974).

V. Predictions of the Impact of Heavy Metal Pollution of Soils

Assessment of heavy metal pollution by chemical analysis can be approached in three ways: first, by assessing the degree of soil or crop pollution; second, by assessing its geographical extent; and finally, by predicting the likely effect on plant and animal health.

A. Degree of Soil Pollution

Measurement of the degree of soil pollution only requires analysis of the total concentration and comparison with an appropriate reference soil, analyzed in the same way which represents the normal "background" situation. Assessment of soil pollution can be made both in terms of time and/or location. Background levels are discussed later. Total analyses of heavy metals do not generally provide any guide to the biological effectiveness or mobility of the metal under study but do define unequivocally the total metal load of a polluted soil. For more useful data on the degree of soil pollution, the total concentration may be divided into different chemical components. The more informative approach is to identify nonlabile and labile forms, the latter having the ability to react to soil conditions. The required procedures could involve successive extractions or a single chemical extraction, provided that conditions of reagent concentration and reaction time are sufficient to totally remove all labile species (Clayton and Tiller, 1979) associated with the soil solid phases. Procedures that only partially remove labile metals from soils are not suitable for defining degree of soil pollution but may usefully predict uptake by crops (see following).

B. Prediction of Uptake of Metals by Plants by Means of Soil Analysis

Soil analysis has been widely used to predict whether a plant's micronutrient metal status lies in the deficient, normal, or toxic range, and has provided a basis of soil amendment, especially for fertilization. The same principal is applied to uptakes of toxic heavy metals.

It is generally accepted that the prediction of trace metal uptake by plants is best achieved by analysis of soil solution or something closely related to it (Kabata-Pendias and Pendias, 1984). The difficulty with this approach lies with the need to separate the appropriate solution, to know which particular ionic species to measure (Sposito and Bingham, 1981), and to analyze the low concentrations involved. Despite the problems, prediction using the soil solution approach holds the best prospect for success. The conventional approach, however, has been to use chemical extractants which, it was hoped, defined a particular chemical form [the chemical pool approach of Viets (1962)] whose relationship to crop composition or fertilizer response could be described statistically. This approach has proved a useful guide to soil and crop management practice. Solutions of dilute acids, complexing agents, and salts have been commonly used for soil testing of metals in both deficient and toxic situations. Soil test predictability has been improved by the application of multivariate statistical analysis using covariant factors that are most likely to affect soil solution composition in aerated soils, for example, pH and complexation. Inertia and lack of convenient alternatives for routine analysis will ensure that these traditional tests continue well into the future.

The complementary approach to trace metal availability and one amenable to theoretical treatment, emphasized the chemical interactions between solid and solution phase, and the factors influencing the equilibria involved (the Q-I studies). These equilibria control both the concentrations in soil solution near the root–soil interface and also the rate of supply to the interface.

Unfortunately a large number of soil tests and their variants are in common use notwithstanding the rather marginal advantage some have over others. Of course many were developed for local use without consideration of the later international applications by others. Procedures developed for calcareous soils [e.g., diethylene triamine pentacetic acid (DTPA); Lindsay and Norvell (1978)] are used on acid sands, and dilute acids suited for acidic-neutral soils are sometimes used on calcareous soils! A bigger hazard arises from the numerous variations in procedure, sometimes in one country, of the commonly used soil tests. Differences in reagent concentration, pH, soil:solution ratio, energy of agitation, etc., which are often unspecified, makes interlaboratory laboratory comparison difficult and leads to a poor-quality data base. This shortcoming, which hinders progress and international cooperation in research and monitoring, is particularly important in the field of pollution which has no respect for international boundaries.

C. Prediction by Plant Analysis

Plant analysis gives an indication of the available heavy metals in soils as well as defining the plant's own metal load, which is important when the plant is ingested as food. For many researchers, the analysis of plant tissue is the preferable approach when assessing the soil's status of nutrient and toxic elements: according to this view, the plant is the only acceptable judge of the soil's ability to transmit the elements under consideration.

However, metal uptake may vary appreciably between genotypes and among different plant parts. Metal status is also influenced by plant maturity, by environmental factors of light, temperature, soil moisture, and rooting depth, and also by interactions with other metal pollutants and nutrients.

D. Measurement of Regional Variations in Heavy Metal Status of Soils

Mineral exploration has made good use of geochemical surveys based on analyses of rocks, soils, and plants. Similar approaches have been used to demarcate areas of varying trace element status; for example, DSIR (1967) in New Zealand and Kubota and Allaway (1972) in the United States. High sampling densities, which are very costly, are required for such soil surveys to ensure meaningful predictions, especially in regions of complex soil patterns and geology. Normal broad-scale soil maps with boundaries determined by morphological properties cannot be confidently used for the extrapolation of trace metal status of soils, except in the comparison of quite contrasting soil types and parent materials. Studies of the ranges of concentrations of trace metals in numerous soil groups of Australian soils (Tiller, 1983) showed appreciable differences between those representing extremes of weathering conditions and parent material, for example, black earths (Vertisols) compared to podzols (Spodosols), but most soil groups had similar ranges of total and extractable trace metal concentration. Pedological surveys of landscapes with less mature soils offer better prospects of extrapolating trace metal data especially if the key characteristics determining the soil boundaries are covariant with trace metal provenance and/or mobility.

Thornton and Webb (1980) have reviewed the methods of mapping the regional distribution of trace metals with special reference to reconnaissance surveys of heavy metals by the analysis of stream sediments, which provide a kind of composite sample of the soils of a drainage basin. This survey technique was established for mineral exploration but has proved useful in the recognition of regional variations in trace and toxic elements. The resulting maps provide rational explanation for noted regional incidences of animal, human, and plant nutritional problems, including the highlighting of areas subjected to historic or current mining and industrial activity. The regional geochemical maps have limited predictive value at the local level but can highlight areas deserving more detailed survey work (Thornton, 1981).

E. Background Levels of Heavy Metal Pollution

Nriagu (1979) has estimated that about 40 million tonnes of heavy metals have been dispersed atmospherically because of man's activities over some thousands of years. Deposition will have been uneven, with more in the northern than the southern hemisphere, and more near industrial centers as compared to rural areas. The resultant global smear of pollution means that pristine (true background) conditions hardly exist anywhere; that is, the concept of a pristine state with reference to heavy metals needs to be applied in a relative sense.

Background values in environmental studies usually refer to a framework of time rather than concentration. Background levels commonly provide a reference point for comparison, for example, before and after the establishment of a factory, the introduction of a new fertilizer practice or a new gasoline additive, etc. Establishment of "background" values for soils and plants is more difficult in industrialized societies of long standing. The overlapping aureoles of aerial contamination from various sources, sometimes from neighboring countries, that are superimposed on centuries of anthropogenic manipulation of the landscape make it impossible to establish background levels of the pristine state. On one hand, we could argue that such values are only of academic interest. On the other hand, the so-called background levels established in an already contaminated area may provide a misleading or false reference point, resulting in the underestimation of the health risk of new industrial or agricultural enterprises or practices. We need to understand better the regional and local variations in natural soil background values.

VI. Guidelines for Regulatory Control of Pollution of Agricultural Land

Increasing concern about heavy metal pollution of soils and crops in relation to human health has led to, or is leading to, the establishment of guidelines for the protection of present and future populations from the adverse effects of pollution. Most of the emphasis to date has been on the conditions of disposal of sewage sludge, but the principles involved could apply to other metal-containing materials deposited on agricultural land.

A. Experimental Approaches

Several difficulties have been recognized. Attempts have been made to generalize the effects of mixtures of heavy metals in terms of a "zinc equivalent" by factorizing supposed relative toxicities of metals (Chumbley, 1971). Recognition that relative toxicity depends on the soil and crop, and that effects may not be additive, has influenced the general acceptance of this approach.

1. Need for Sound Experimental Basis

Pot versus field experiments: Responses of plants to toxic metals in pot experiments may greatly exceed uptake under field conditions (de Vries and Tiller, 1978; Chang and Page, 1979). De Vries and Tiller (1978) compared the uptake by onions (*Allium cepa*) and lettuces (*Lactuca sativa*) grown on the same sludge-treated soil under three experimental conditions: pots with 16 kg soil; large concrete containers with 1 m^3 soil; and on plots in a farmer's field. Figure 10 shows that Cd concentrations and the response curves for plants grown in pots, even with 16 kg of soil, provide no guide to the likely risks of growing vegetables

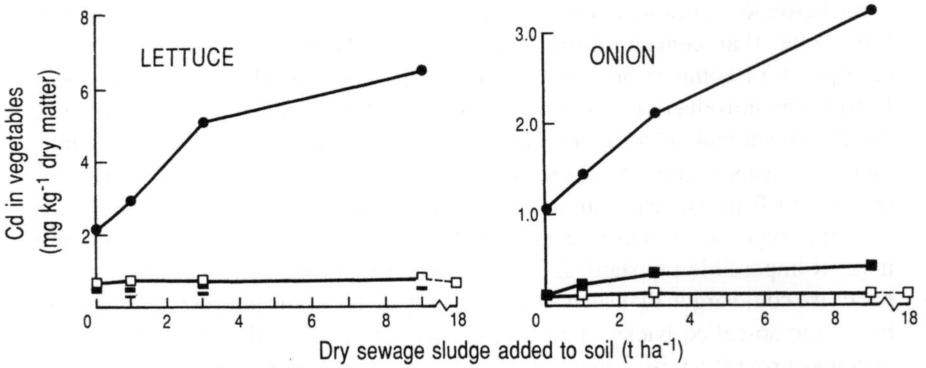

Figure 10. Effect of different experimental conditions on the uptake of Cd from sewage sludge-treated soil (field, □; miniplot, ■; pots in glasshouse, ●) by lettuce and onion (from de Vries and Tiller, 1978).

under field conditions because of differences in environmental conditions and rooting pattern.

Dose Rates: Dose rates in experiments should also realistically cover the normal range of field applications of sludge, etc.

Chemical Form of Heavy Metal: Use of soluble forms of toxic metals may not simulate reactions of metals in slag, sludge or fertilizer mixtures. The validity of this approach, if used, will depend on the pollutant source under evaluation and soil conditions. Tiller and de Vries (unpublished data, reported by Tiller, 1979) showed in isotopic dilution studies that Cd and Zn in Adelaide sewage sludges had the same solubility as chloride salts when either was added to neutral sandy loam soil. This contrasts with other investigations (Korcak and Fanning, 1978), in which it was found that Cd, Zn, Cu, and Ni in sludge were much less soluble than a simple inorganic salt. Such contrasts must reflect the variation of chemical form of heavy metal in different sewage sludges in equilibrium with the particular soil matrix.

2. Evaluation of Long-Term Effects

Evaluation of experimental results after brief equilibration periods may give false indications of long-term effects. Bates et al. (1975) measured the trend in heavy metal concentration of four successive harvests of ryegrass (*Lolium perenne*) during a year. Final concentrations of Cd, Cu, and Zn in ryegrass following a single initial addition of sewage sludge decreased to one half or less of those of the initial harvest, but Ni was unchanged. Regular additions of sludge before each crop did not increase uptake. Coker et al. (1982) showed a 50% decrease in Cd concentration in ryegrass within 2 years of sludge addition with little further change during following years. Hinesly et al. (1979) reported that whereas Cd and Zn in maize grain (*Zea mays*) decreased over 4 years to near-background levels following the last application, concentrations in leaves maintained higher levels. Dowdy et al. (1978) found no change in Cu and Zn contents of beans (*Phaseolus vulgaris*) during several years after sludge addition. Chaney and Giordano (1977) quoted several studies showing no increase in Cd contents of fodder crops grown for many years on sludge-treated soils.

There is general agreement that applications of different amounts of metal-enriched sludge to soils have an additive effect on metal loads of crops when applied on one occasion but much less and sometimes not at all, when the additions are separated by appreciable time intervals. The evidence suggests that heavy metal availability to crops only decreases appreciably within several years, at most. This assumed fixation of heavy metals in sludge-treated soils within a short time span must have been an important consideration in the development of some national guidelines for sludge disposal following sludge application. However, the suggestion (Hinelsy et al., 1977) that only the last sludge addition is of significance to crop quality may well be an oversimplification that is very site and crop dependent and thus a risky basis for long-term disposal of wastes.

Possible longer term changes in availability to crops are difficult to predict because of the complex interacting factors involved, namely:

1. Cd, for example, is known to have strong affinity for organic as well as mineral surfaces, especially iron oxidic, with the organic surfaces favored at lower pH (Tiller et al., 1984). The extent, nature of, and competition between, different surfaces may change with time.
2. Reactions of organic constituents of sludge or their microbial breakdown products with heavy metals do not necessarily provide the fixation sink proposed by Leeper (1978) and others, but rather may enhance, in the shorter term, heavy metal transport and uptake by plants (Hinesly et al., 1977). The sludge constituents responsible for binding of heavy metals, especially for the mobility/uptake function, may be those most susceptible to microbial degradation.
3. Soil pH may be slowly changed by agricultural practice, including disposal of sewage sludge and other wastes, and by atmospheric fallout, including acidic rain, and thus affect equilibria affecting solubility/availability of heavy metals in soils with time.
4. Some heavy metals may be "lost" from the surface layers by erosion and/or dilution by cultivation. Some apparent losses may be the consequence of use of chemical procedures that inadequately recognize transformation to less insoluble forms.
5. Assessment of the long-term availability of heavy metals and their impact on crop quality depends on the plant part analyzed; impeded translocation of some heavy metals to seeds rather than leaves, for example, may disguise the possible impact on pollution or more susceptible plants or plant parts.

To sum up, guidelines controlling the application of sewage sludge and other wastes must take into account not only the important factors of composition, application rate, and thus the limitation of maximum levels of heavy metals appropriate to existing soils/management conditions, but also the possible long-term changes in availability to plants and mobility in the soil/hydrological system. These changes may be natural, but can also be strongly affected by anthropogenic influences related to either farm management or to urban-industrial pollution.

Farm decisions that affect fertilizer and irrigation practices and crop rotations can result in gradual changes in organic matter, pH, and possible redox conditions, and thus prejudice at some later stage the status quo of heavy metal availability to crops. Gradually declining pH values over several decades, for example, as occur on soils associated with permanent leguminous pastures in Australia (Helyar, 1976), result in widespread aluminum toxicity but could lead imperceptibly to additional hazards if the soils were polluted with heavy metals such as cadmium. Urban-industrial activities well removed from farming communities can also have a major impact on the biological effectiveness of heavy metals in soils. The widespread dispersal of acidic industrial gases into the atmosphere and the resultant acid rain could dramatically increase the threat of heavy metal toxicity on weakly buffered soils unless the soils are limed regularly.

B. Legislative Approaches

Guidelines for disposal of sewage sludge have been developed in several ways (CEC, 1982), namely:

1. By restricting the heavy metal concentrations in sludge being applied to agricultural land.
2. By restricting the amount of pollutant metal that can be added over a prescribed period. In practice this in achieved by controlling both the maximum permissible concentrations of metals in sludge and the amounts of sludge that can be added. An important benefit of this approach is the incentive to improve controls of industrial effluents.
3. By defining a maximum permissible concentration of heavy metals in agricultural soils.
4. By controlling the heavy metal contents of foodstuffs.
5. By combinations of these, combined with restrictions related to soil characteristics, land use and crop type, and the metals controlled.

Many countries agree on the need for guidelines to control the pollution of agricultural soils, but the great range of approaches and permissible limits is surprising. Some complications of these disparate regulations have been reviewed by Webber et al. (1983) and Purves (1983). Maximum permissible Cd concentrations in sludge varied from 8 to 30 mg kg^{-1}, and maximum permissible loading of soil with Cd in sludge varied from 0.1 to 20 kg ha^{-1} (CEC, 1982). Although one recognizes the inertia of parochial scientific approaches and traditions, it is surprising that the huge available literature on soil pollution has not resulted in greater similarity in approach and better consensus of recommendations in different countries. Political and nonscientific community inputs are likely to have played a significant role. On the other hand, this disparity of recommended limits may reflect lack of confidence in the scientific basis, especially of the relationship between the concentrations of metals in soil and uptake by plants, the question of whether or not availability of metals in polluted soils changes with time, and the significance of the critical concentration of heavy metals in food plants in relation to suggested dietary limits.

VII. Final Comment

The health and sustenance of future generations demands that our soil resource be protected against a slow, insidious poisoning by heavy metals released by urban-industrial, mining, and agricultural activities; this is a challenge for both our legislators and their scientific advisors. Soil scientists can, and should, continue to play their significant role in investigations leading to further development and refinement of national guidelines for the disposal and use of toxic materials on agricultural land. Soil analysis, sample collection and preparation carried out in a prescribed adequate manner can provide a reliable estimate of the loading of soil by heavy metals from most pollution sources, and can also reflect unusual variations in background concentrations on which the pollution is superimposed. The usefulness of such soil data can be increased by coupling them with critical soil factors that influence metal availability, or covariants of them suitable for use in regulations.

An example is the use of pH and cation-exchange capacity in the United States regulations (EPA, 1979). The normal cation-exchange reactions that take place in soils may have limited direct relevance to the critical reactions of heavy metals that affect availability and long-term retention. However, the cation-exchange capacity will often be covariant with other soil variables, such as the area of reactive surfaces, fine clay content etc., which may be more relevant. These critical factors could differ in emphasis in different regions. Discrimination between insoluble and potentially labile forms of metal in the soil would also make soil analysis a more useful basis of soil pollution guidelines, especially if applied internationally.

An important challenge remaining for scientists is the prediction of possible long-term changes in the hazard of soils polluted by heavy metals arising from variations in soil conditions. Regulations should therefore ensure that polluted soils, especially those receiving maximum allowable metal loadings, are safeguarded against future management decisions likely to mobilize accumulated metals into crops or groundwaters. Farmers will not wish their land use to be controlled but we should, through guidelines based on research, avoid Leeper's (1978) "possible day of reckoning."

References

Baechle, H., and F. Wolstein. 1984. *Cadmium Compounds in Mineral Fertilizers*. The Fertilizer Society, London.

Bates, T.E.,A. Haq, U.K. Soon, and J.A. Moyer. 1975. Uptake of metals from sewage sludge-amended soils. *Heavy Metals Environ. Int. Conf.*, 2nd 1:403–416.

Brownell, P.F., and C.J. Crossland. 1972. The requirement for Na as a micronutrient by species having the C_4-dicarboxylic photosynthesis pathway. *Plant Physiol.* 49:794–797.

Brümmer, G., K.G. Tiller, U. Herms, and P.M. Clayton. 1983.Adsorption-desorption and/or precipitation-dissolution processes of zinc in soils. *Geoderma* 31:337–354.

Cartwright, B., R. H. Merry, and K.G. Tiller. 1977. Heavy metal contamination of soils around a lead smelter at Port Pirie, South Australia. *Aust. J. Soil Res.* 15:69–81.

C.E.C. 1982. Proposal for a Council Directive on the use of sewage sludge in agriculture. *Comm. Eur. Communities Eur. Rep.* C264:3–8.

Chamberlain, A.C. 1983. Fallout of lead and uptake by crops. *Atmos. Environ.* 17:693–706.

Chaney, R.L., and P.M. Giordano. 1977. Microelements as related to plant deficiencies and toxicities. In: *Soils for management of organic wastes and waste waters*, pp. 235–279. American Society of Agronomy, Madison, Wisconsin.

Chang, A.C., and A.L. Page. 1979. Effects of repeated soil application of sludges on the Cd and Zn levels of affected vegetable crops. *Final Report*, Wastewater Solids Management Program, Los Angeles/Orange County Metropolitan Area. University of California, Riverside, California.

Chumbley, G.C. 1971. Maximum permissible levels of metals in sewage applied to agricultural land. *ADAS Adv. Pap. No. 10.* MAFF, London.

Clayton, P.M., and K.G. Tiller. 1979. A chemical method for the determination of the heavy metal content of soils in environmental studies. *CSIRO Aust. Div. Soils Tech. Pap. No. 41.*

Coker, E.G., Davis, J.E. Hall, and C.N. Carlton-Smith. 1982. The use of sewage sludge in land reclamation. *Water Research Centre, Stevenage, Tech. Rep. TR 183.*

Davis, R.D. 1980. Control of contamination problems in the treatment and dispersal of sewage sludges. *Water Research Centre, Stevenage, Tech. Rep. TR156.*

Davis, R.D. 1983. Sewage sludge utilization on land: recent developments in research on cadmium. *Heavy Metals Environ. Int. Conf.*, 4th 1:330–341.

Davis, R.D., and C.N. Carlton-Smith. 1980. Crops as indicators of the significance of contamination of soils by heavy metals. *Water Research Centre, Stevenage, Tech. Rep. TR140.*

de Vries, M.P.C. 1983. Investigations on twenty Australian sewage sludges—their evaluation by chemical analysis. *Fert. Res.* 4:75–87.

de Vries, M.P.C., and K.G. Tiller. 1978. Sewage sludge as a soil amendment with special reference to Cd, Cu, Mn, Ni, Pb and Zn—Comparison of results from experiments conducted inside and outside a glasshouse. *Environ. Pollut.* 16:231–240.

de Vries, M.P.C., K.G. Tiller and L.R. Spouncer. 1975. Environmental pollution of the Port Pirie region. 2. Concentrations of cadmium, lead and zinc in plants grown under glasshouse conditions on contaminated soils. *CSIRO Aust. Div. Soils Rep. No. 7.*

Domsch, K.H. 1984. Effects of pesticides and heavy metals on biological processes in soil. *Plant Soil* 76:367–378.

Dowdy, R.H., W.E. Larson, J.M. Titrud, and J.J. Latterell. 1978. Growth and metal uptake of snap beans grown on sewage sludge-amended soil: a four-year field study. *J. Environ. Qual.* 7:252–257.

DSIR. 1967. *Soil Bureau Atlas*. DSIR, Wellington, New Zealand.

Emmerich, W.E., L.J. Lund, A.L. Page, and A.C. Chang. 1982. Movement of heavy metals in sewage-sludge treated soils. *J. Environ. Qual.* 11:174–178.

Encyclopedia Britannica. 1980. *The New Encyclopedia Britannica*, Vols. 3,4,10, 13, and 19. Encyclopedia Britannica, Inc., Chicago, Illinois.

E.P.A. (U.S. Environmental Protection Agency). 1979. Critieria for classification of solid waste disposal facilities and practices. *Fed. Reg.* 44:53438–53468.

Freedman, B., and T.C. Hutchinson. 1981. Sources of metal and elemental contamination of terrestrial environments. In: N.W. Lepp (ed.), *Effect of Heavy Metal Pollution on Plants*, Vol. 2. Applied Science, London.

Gerth, J., and G. Brümmer. 1983. Adsorption und Festlegung von Nickel, Zink, und Cadmium durch Goethit (α-FeOOH). *Fresenius Z. Anal. Chem.* 316:616–620.

Graham, D., and S.M. Kalman. 1974. Lead in forage grass from a suburban area in Northern California. *Environ. Poll.* 7:209–215.

Gulson, B.L., K.G. Tiller. K.J. Mizon, and R.H. Merry. 1981. Use of lead isotopes in soils to identify the source of lead contamination near Adelaide, South Australia. *Environ. Sci. Technol.* 15:691-696.

Gunnarsson, O. 1983. Heavy metals in fertilizers. Do they cause environmental and health problems? *Fert. Agric.* No. 85:27–42.

Hannam, R.J., and D.J. Reuter. 1977. The occurrence of steely wool in South Australia 1972–1975. *Agric. Rec. (S. Aust.)* 4:26–29.

Healy, W.B. 1973. Nutritional aspects of soil ingestion by grazing animals. In: G.W. Butler and R.W. Bailey (eds.), *Chemistry and Biochemistry of herbage*, Vol. 1, pp. 567–588. Academic Press, London.

Helyar, K.R. 1976. Nitrogen cycling and soil acidification. *J. Aust. Inst. Agric. Sci.* 42:217–221.

Hinesly, T.D., R.L. Jones, E.L. Ziegler, and J.J. Tyler. 1977. Effects of annual and accumulative applications of sewage sludge on assimilation of zinc and cadmium by corn (*Zea mays* L.) *Environ. Sci. Technol.* 11:182–188.

Hinesly, T.D., E.L. Ziegler, and G.L. Barrett. 1979. Residual effects of irrigating corn with digested sewage sludge. *J. Environ. Qual.* 8:35–38.

Hirst J.M., H.H. Le Riche, and C.L. Bascomb. 1961. Copper accumulation in the soils of apple orchards near Wisbech. *Plant Pathol.* 10:105–108.

Hodgson, J.F., H.R. Geering, and W.A. Norvell. 1965. Micronutrient cation complexes in soil solution: Partition between complexed and uncomplexed forms by solvent extraction. *Soil Sci. Soc. Am. Proc.* 29:665–669.

Hughes, M.K., N.W. Lepp, and D.A. Phipps. 1980. Aerial heavy metal pollution and terrestrial ecosystems. *Adv. Ecol. Res.* 11:217–327.

Hutchinson, T.C. 1981. Nickel. In: N.W. Lepp (ed.), *Effect of Heavy Metal Pollution on Plants*, Vol. 1, pp. 171–212. Applied Science, London.

Kabata-Pendias. A., and H. Pendias. 1984. *Trace Elements in Soils and Plants*. CRC Press, Baton Toca, Florida.

Klein, L.A., M. Lang, N. Nash, and S.L. Kirschner. 1974. Sources of metal in New York City wastewater. *J. Water Pollut. Control Fed.* 46:2653–2662.

Kofoed, A.D. 1984. Optimum use of sludge in agriculture, In: *Utilization of Sewage Sludge on Land. Comm. Eur. Communities Eur. Rep.* 8822.

Koh, T.-S., and G.J. Judson. 1986. Trace-elements in sheep grazing near a lead-zinc smelting complex at Port Pirie, South Australia. *Bull. Environ. Contam. Toxicol.* 37:87–95.

Korcak, R.F., and D.S. Fanning. 1978. Extractability of cadmium, copper, nickel, and zinc by double acid versus DTPA and plant content at excessive soil levels. *J. Environ. Qual.* 7:506–512.

Kubota, J., and W.H. Allaway. 1972. Geographic distribution of trace element problems In: J.J. Mortvedt, P.M. Giordano, and W.L. Lindsay (eds.) *Micronutrients in Agriculture*, pp. 525–554. Soil Science Society of America, Madison, Wisconsin.
Küster, E., and I. Grün. 1984. Cadmium und Bodenmikroorganismen. *Angew. Botanik* 58:31–38.
Låg, J. 1985. Soil pollution by cadmium from incineration plants. *Ambio* 14:356.
Lee, K.E. 1985. *Earthworms. Their Ecology and Relationship with Soils and Land Use.* Academic Press, New York.
Leeper, G.W. 1972. Reactions of heavy metals with soils with special regard to their application in sewage wastes. U.S. Dept. of Army Corps of Engineers, Contract No. DACW73-73-C-0026. 70pp.
Leeper G.W. 1978. *Managing the Heavy Metals on Land.* Marcel Dekker, New York.
Lindsay, W.L., and W.A. Norvell. 1978. Development of a DTPA soil test for zinc, iron, manganese and copper. *Soil Sci. Soc. Am. J.* 42:421–428.
McMichael, A.J., G.V. Vimpani, E.F. Robertson, D.D. Clark, and D.A. Baghurst. 1986. The Port Pirie cohort study: maternity blood lead and pregnancy outcome. *J. Epidemiol. Community Health* 40:18–25.
Merry, R.H., K.G. Tiller. 1978. The contamination of pasture by a lead smelter in a semi-arid environment. *Aust. J. Exp. Al ric. Anim. Husb.* 18:89–96.
Merry, R.H., K.G. Tiller, and A.M. Alston. 1983. Accumulation of copper, lead, and arsenic in some Australian orchard soils. *Aust. J. Soil Res.* 21:549–561.
Merry, R.H., K.G. Tiller, M.P.C. de Vries, and B. Cartwright. 1981. Contamination of wheat crops around a lead-zinc smelter. *Environ. Pollut. Ser. B. Chem. Phys.* 2:37–48.
Mochizuki, H., S. Chiba, S. Hanada, and H. Saitoh. 1975. Ecological study on the apple orchards contaminated by inorganic agricultural chemicals. *Nippon Dojo-Hiryogaku Zasshi* 46:45–50.
Mortvedt, J.J. 1987. Cadmium levels in soils and plants from some long-term soil fertility experiments in the United States of America. *J. Environ. Qual.* 16:137–142.
Müller, G. 1981. Heavy metals and other pollutants in the environment: A chronology based on the analysis of dated sediments. *Heavy Metals Environ. Int. Conf., 3rd*: 12–17.
Murozumi, M., T.J. Chow, and C. Patterson. 1969. Chemical concentrations of pollutant lead aerosols, terrestrial dusts, and sea salts in Greenland and Antarctic snow strata. *Geochim. Cosmochim Acta* 33:1247–1294.
Nicholas, D.J.D. 1975. The functions of trace elements in plants. In: D.J.D. Nicholas and A.R. Egan (eds.), *Trace Elements in Soil-Plant-Animal Systems*, pp. 181–198. Academic Press, London.
Nriagu, J.O. 1979. Global inventory of natural and anthropogenic emissions of trace elements to the atmosphere. *Nature (London)* 279:409–411.
Page, A.L. 1974. Fate and effects of trace elements in sewage sludge when applied to agricultural lands. EPA-670/2-774-005, U.S. Environmental Protection Agency, Cincinnati, Ohio.
Page, A.L., and T.J. Ganje. 1970. Accumulation of lead in soils for regions of high and low motor vehicle traffic density. *Environ. Sci. Technol.* 4:140–142.
Page, A.L., A.A. Elseewi, and I.R. Straughan. 1979. Physical and chemical properties of fly ash from coal-fired power plants with reference to environmental impacts. *Residue Rev.* 71:83–114.
Page A.L., F.T. Bingham, and A.C. Chang. 1981. Cadmium. In: N.W. Lepp (ed.), *Effect of Heavy Metal pollution on plants*, Vol. 1, pp. 77–110. Applied Science. London.

Phipps, D.A 1981. Chemistry and biochemistry of trace elements in biological systems. In: N.W. Lepp (ed.), *Effects of Heavy Metal Pollution on Plants,* Vol. 1, pp. 1–54. Applied Science, London.

Purves, D. 1977. *Trace-Element Contamination of the Environment.* Elsevier, Amsterdam.

Purves, D. 1983. EC Directive on the use of sewage sludge in agriculture–environmental implications. *Heavy Metals Environ. Int. Conf. 4th* 1:342–345.

Reiter, E.R., T. Henmi, and P.C. Katen. 1977. Modeling atmospheric transport. In: *Lead in the Environment*, pp. 73–92. National Science Foundation, Washington, D.C.

Reuss, J.O., H.L. Dooley, and W. Griffis. 1978. Uptake of cadmium from phosphate fertilizers by peas, radishes and lettuce. *J. Environ. Qual.* 7:128–133.

Schroeder, H.A., and J.J. Balassa. 1963. Cadmium: uptake by vegetables from superphosphate in soil. *Science* 140:819–820.

Smilde, K.W., and B. van Luit. 1983. The effect of phosphate fertilizer cadmium on cadmium in soils and crops. *Inst. Bodemvruchtbaarheid Haren-Gr. Rapp.* 6-83:1–17.

Sposito, G. 1983. The chemical forms of trace elements in soils. In: *Applied Environmental Geochemistry.* Academic Press, New York.

Sposito, G., and F.T. Bingham. 1981. Computer modelling of trace metal speciation in soil solutions: correlation with trace metal uptake by higher plants. *J. Plant Nutr.* 3:35–49.

Stenström, T., and Vahter, M. 1974. Cadmium and lead in Swedish commercial fertilizers. *Ambio* 3:91–92.

Suttle, N.F. 1975. Trace element interactions in animals. In: D.J.D. Nicholas and A.R.Egan (eds.) *Trace Elements in Soil-Plant-Animal Systems*, pp. 271–290. Academic Press, New York.

Swaine, D.L. 1962. The trace-element content of fertilizers. *Commonw. Bur. Soils Tech. Commun. No.* 52.

Takijima, Y., and F. Katsumi. 1973. Cadmium contamination of soils and rice plants caused by zinc mining. *Soil Sci. Plant Nutr.* 19:29–38.

Thornton, I. 1981. Geochemical aspects of the distribution and forms of heavy metals in soils. In: N.W. Lepp (ed.), *Effect of Heavy Metal Pollution on Plants*, Vol. 2, pp. 1–34. Applied Sciences. London.

Thornton, I., and J.S. Webb. 1980. Regional distribution of trace element problems in Great Britain. In: B.E. Davies (ed.), *Applied Soil Trace Elements*, pp. 1–34 Wiley, New York.

Tiller, K.G. 1963. Weathering and soil formation on dolerite in Tasmainia with particular reference to several trace elements. *Aust. J. Soil Res.* 1:74–90.

Tiller, K.G. 1979. Applications of isotopes to micronutrient studies. *IAEA Proc. Ser. Colombo 1978*, pp. 359–372. IAEA, Vienna.

Tiller, K.G. 1983. Micronutrients. In: *Soils: An Australian Viewpoint*, Chap. 25. CSIRO, Melbourne/Academic Press, London.

Tiller, K.G., B. Cartwright, M.P.C. de Vries, R.H. Merry, and L.R. Spouncer. 1975. Environmental pollution of the Port Pirie region. 1. Accumulation of metals in wheat grain and vegetables grown on the coastal plain. *CSIRO Aust. Div. Soils Rep. No. 6.*

Tiller, K.G., M.P.C. de Vries, L.H. Smith, and B. Zarcinas. 1976. Environmental pollution of the Port Pirie region. 3. Metal contamination of home gardens in the city and their vegetable produce. *CSIRO Aust. Div. Soils. Rep. No. 15.*

Tiller, K.G., and M.P.C. de Vries. 1977. Contamination of soils and vegetables near the lead-zinc smelter, Port Pirie, by cadmium, lead and zinc. *Search* 8:78–79.

Tiller, K.G., and R.H. Merry. 1982. Copper pollution of agricultural soils. In: J.F. Loneragan, A.D. Robson, and R.D. Graham (eds.) *Copper in Soils and Plants*, pp. 119–140. Academic Press, London.

Tiller, K.G., J. Gerth, and G. Brümmer. 1984. The relative affinities of Cd, Ni and Zn for different soil clay fractions and goethite. *Geoderma* 34:17–35.

Tiller, K.G., L.H. Smith, R.H. Merry, and P.M. Clayton. 1987. The dispersal of automotive lead from metropolitan Adelaide into adjacent rural areas. *Aust. J. Soil Sci.* 25:155–166.

Tyler, G. 1972. Heavy metals pollute Nature, may reduce productivity. *Ambio* 1:52–59.

Tyler, G. 1981. Heavy metals in soil biology and biochemistry. In: E.A. Paul and J.N. Ladd (eds.), *Soil Biochemistry*, pp. 372–414 Marcel Dekker, New York.

Underwood, E.J. 1977. *Trace Elements in Human and Animal Nutrition*. 4th Ed. Academic Press. London.

Unwin, R.J. 1981. The application of copper in sewage sludge and pig manure to agricultural land in England and Wales. In: *Copper in Animal Wastes and Sewage Sludge. Comm. Eur. Communities Eur. Rep. 7196.*

van Driel, W., H.N. Kerdijk, and W. Salomons. 1984. Use and disposal of contaminated dredged material. *Land Water Int.* 53:13–18.

Viets, F.G. 1962. Chemistry and availability of micronutrients in soils. *Agric. Food Chem.* 10:174–178.

Vinogradov, A.P. 1959. *The Geochemistry of Rare and Dispersed Chemical Elements in Soils*. Consultants Bureau, New York.

Walker, C.D., R.G. Graham, J.T. Madison, E.E. Cary, and R.M. Welch. 1985. Effects of Ni deficiency on some nitrogen metabolites in cowpeas (*Vigna unguiculata* L. Walp). *Plant Physiol.* 79:474–79.

Webber, M.D., A. Kloke, and J.C. Tjell. 1983. A review of current sludge use guidelines for the control of heavy metal contamination in soils. In: *Processing and Use of Sewage Sludge. Comm. Eur. Communities Eur. Rep. 9129.*

Wheeler, G.L., and G.L. Rolfe. 1979. The relationship between daily traffic volume and the distribution of lead in roadside soil and vegetation. *Environ. Pollut.* 18:265–274.

Williams. C.H., and D.J. David. 1973. The effect of superphosphate on the cadmium content of soils and plants. *Aust. J. Soil Res.* 11:43–56.

Zhu, Q-Q. and Z. Liu. 1986. Status of rare earth elements in soils. In: *Current Progress in Soil Science in People's Republic of China*, pp. 335–339. Soil Sci. Soc China.

The Use of Extractants in Studies on Trace Metals in Soils, Sewage Sludges, and Sludge-Treated Soils

P.H.T. Beckett

I. Introduction 144
A. The Problem 144
B. This Study 145
II. Forms of Combination of Trace Metals in Sewage Sludge, Sludged Soils, or Soils 146
III. Survey of Extractants 149
A. Organic Solvents 149
B. Water 150
C. Simple Salts 150
D. Less Simple Salts 151
E. Acetic Acid 152
F. Acetates 153
G. Oxalic acid and Oxalates 153
H. Sodium Citrate 154
I. Fluorides 154
J. Acetyl Acetone 155
K. EDTA 155
L. DTPA 157
M. Sodium/Potassium Pyrophosphate 157
N. Acid Ammonium Oxalate (and Ammonium Tartrate) 160
O. Hydroxylamine Hydrochloride (and Hydroquinone) 162
P. Sodium Dithionite 163
Q. Sodium Hypochlorite (and Hypobromite) 164
R. Hydrogen Peroxide 165
S. Sodium Hydroxide 165
T. Sodium Carbonate 166
U. Nitric Acid 166
V. Hydrochloric Acid 167
W. Strong Acids 167
IV. Discussion 168
References 171

Advances in Soil Science, Volume 9

I. Introduction

A. The Problem

When the main problems of trace metals in the soil were deficiencies, it was usually sufficient to know how much of an essential trace metal was "available" to an extractant, usually acetic acid, ethylene diamine tetracetic acid (EDTA), diethylene triamine pentacetic acid (DTPA), etc. (e.g., Lindsay and Norvell, 1969; Viets and Lindsay, 1973; Ministry of Agriculture, Fisheries and Food, 1981). This is fortunate because the soil chemistry of trace metals is rarely simple, and it is particularly obscure at deficiency levels, when even the least unavailable trace metal ions are held on "specific sites" whose character is difficult to establish or define, rather than in identifiable salts (Lindsay, 1979).

On the other hand, environmental pollution by trace metals in sewage sludges and other organic wastes is a problem because large applications of trace metals can saturate the "specific sites" that might hold them in relatively unavailable forms, and still leave an excess to be held in more reactive and potentially toxic forms. In this case the protection of land from pollutants, and the reclamation of polluted land, may require some understanding of the forms of combination of the trace metals in pollutants and their chemistry in soil. Indeed, it may be impossible to predict the long-term fate of trace metals from pollutants without knowing their principal forms of combination and the likely transformations of these combinations. The latter is particularly relevant when trace metals are applied to land as contaminants of digested sewage sludge, in which they may be present in compounds stable under anaerobic but metastable in aerobic conditions, or held by easily decomposed ligand groups on organic materials.

Sewage sludge is a complex mixture of organic and inorganic materials and compounds (Baldwin et al., 1983). It is not easy to isolate distinct fractions in order to study their trace metals without altering them. A few workers (e.g., Bergmann et al., 1979) have separated mineral fractions, more or less impure, by sedimentation or density separation and then identified discrete compounds of trace metals by x-ray diffraction (XRD) or by scanning electron microscope (SEM) with microprobe (Mattigod et al., 1986) but it may be difficult to apply these techniques to mixtures of colloidal organic matter and inorganic precipitates. In simpler systems it might be possible to demonstrate that a particular trace metal compound is present by recognizing a stable activity product in the equilibrium solution of a soil or pollutant, or it might be possible to distinguish different kinds of exchange surfaces by the forms of their exchange isotherms for trace metals and the effects of pH on these isotherms (Himes and Barber, 1957). Both techniques will be difficult in the case of organic wastes such as sewage sludge in which the equilibrium solution is dominated by organic ligands of uncertain identity.

These problems, and others, have led many workers to employ "specific extractants" to identify the forms of combination of trace metals in sludge or soil, or in geological sediments. Generally three to eight extractants are used in a sequence in which the earlier ones are the least aggressive and the most specific, and subsequent extractants are progressively more destructive and less specific, except that coming late in the sequence there may remain only one or two groups of compounds that they can dissolve. Thus, Stover et al. (1976) compared two sequences:

i. 0.05 *M* $CaCl_2$, 24 h; 2½% (v/v) acetic acid, 24 h; 0.1 *M* $Na_4P_2O_7$, 16 h; (0.175 *M* ammonium oxalate and 0.1 *M* oxalic acid (pH 3.25), 100°C, 2.5 h) × 2.
ii. 1 *M* KNO_3, 16 h; 0.5 *M* KF (pH 6.5), 16 h; 0.1 *M* $Na_4P_2O_7$, 16 h; [0.1 *M* EDTA (pH 6.5), 8 h] × 2; 1 *M* HNO_3,16 h.

Some authors present the results of a sequential extraction in terms of the proportions of each trace metal that were "exchangeable," "soluble," or present as undefined "precipitates" or as sulfides or carbonates, "associated with organic matter," "associated with manganese oxides," or "associated with iron and aluminium oxides," etc. (Neilsen et al., 1986; and many others). Other workers have been less confident. Emmerich et al. (1982) stated "the terminology [of extracted fractions such as "adsorbed," "carbonate," etc.] is not meant to imply that the metal is in only adsorbed, carbonate, etc., form but extractable with that reagent, and "the designated chemical forms extracted by specific reagents . . . are not meant to imply specific forms of the metal whose forms of combination are being investigated but to say that the metal forms are chemically similar and extractable with specific reagents."

Even this may be too optimistic; there is no reason why one extractant should not dissolve two or more compounds that are not chemically similar, or why it must necessarily be able to dissolve chemically similar compounds (e.g., carbonates) of two or more elements (Stover et al., 1976). Miller and McFee (1983) pointed out that "fractionation methods do not resolve solid phases from metals adsorbed or occluded by soil surfaces," and there are other problems. Hickey and Kittrick (1984) concluded that "methods of sequential extraction are operationally defined by the sequence of extractants used": they might better have said that the fractions extracted are operationally defined by the extractants that can or cannot dissolve them.

The situation is further confused by the natural tendency of nonspecialists to adopt published extractant sequences as standard procedures without appreciating the reservations of their proposers. For example, the effects of acid ammonium oxalate extractant have been clearly shown (see below) to depend on the nature and degree of illumination of the reaction mixture (e.g., dark, daylight, or ultraviolet light), yet many publications do not specify which illumination was employed, and some workers at least appear to have worked under the undefined illumination of an open laboratory. Other workers have not recorded the pH or concentration of their EDTA, pyrophosphate, or other extractants.

B. This Study

In view of these uncertainties, this chapter reviews the use of extractants in studies on trace metals in soils, or in soils to which sewage sludge (in particular) or other wastes have been applied. It attempts to summarize which group of compounds the users thought each extractant was dissolving and to generalize their findings. The review opens with a brief summary of the ways in which trace metals may be held in sewage sludge or soils, and closes with some general comments on the extractants used in this field and on how their specificity might be increased.

II. Forms of Combination of Trace Metals in Sewage Sludge, Sludged Soils, or Soils

An anaerobic digester contains a substantial population of anaerobic microorganisms that destroy most of the easily composed organic materials and leave resistant plant residues, fresh biomass, and stable heteropolymers. The reducing conditions give rise to a pH of 7 to 8 and iron(II), manganese(II), sulphide and bicarbonate, and possibly copper(I) ions. Usually these and the dominant calcium and magnesium cations form carbonate, sulfide, phosphate, or hydroxide precipitates in the sludge. Substantial substitution of trace metals and other minor elements in their crystal lattices is likely. Also, trace metals are adsorbed onto the imperfectly crystalline surfaces of non-aged precipitates, and may become occluded beneath later precipitates.

After it is removed from the digester, the sludge becomes aerobic and various changes ensue. In particular, sulfide precipitates of trace metals become oxidized and release their cations. Iron(II) and manganese(II) compounds, particularly iron sulfides, revert to oxides with a fall in pH and the release of occluded trace metals. The oxides formed may occur as discrete deposits or as coatings on other materials. The dead anaerobes become the substrate for a new population of aerobes, generating new microbial tissue and releasing various metabolites, including polysaccharide exudates, with new groups of ligand sites.

Thus, the trace metals within sludge, oxidized sludge, or soil may occur in any of the following forms:

1. *Soluble*: as free ions, or as soluble complexes with inorganic anions or organic ligands. Note that chloride and sulfate tend to complex more strongly than nitrate, and soluble organic anions more strongly than either. On the organic ligands there may be some sites that can hold trace metals and simple basic cations with comparable affinity, and others where the former have a strong advantage (Fletcher and Beckett, 1987).
2. *Exchangeable*: held by predominantly electrostatic forces on negatively charged sites on clays, other minerals or organic matter, or on amorphous materials with a low pH of zero charge. By concept, exchangeable cations are isotopically exchangeable and may also be displaced by the basic cations commonly present in soil or sludge solutions. The activation energy of exchange must be low, because this exchange is usually rapid and complete. However, the distinction between electrovalently and covalently bonded cations is not

sharp, and some cations may be taken up rapidly by electrovalent forces and then pass over slowly to a coordinate or covalent combination (see Group 3) from which simple basic cations can displace them only slowly, if at all.

3. *Specific adsorption*: held on sites (usually assumed to be inorganic) at which one element, or the transition elements more generally, are held relatively strongly by predominantly covalent or coordinate forces (Miller et al., 1986a). Ions held in this way are taken up and released more slowly than exchangeable ions, and are more easily displaced by other trace metals or by hydrions than by basic cations (Hodgson, 1963). They often are, but need not be, isotopically exchangeable. Measurements of their ease of replacement sometimes fail to distinguish the effects of a high activation energy (slow exchange) or a high energy of complexation (adverse equilibrium); either or both may apply. The number of specific adsorption sites usually varies with pH, and there may be sites that can hold H_3O^+ or trace metals, but not alkali or alkali earth cations (Himes and Barber, 1957).

   The concept of this group is not precise: it grades into Group 2 and Group 4. It also covers a very heterogeneous range of materials. There will be a small number of specific sites even on clay minerals or other mineral surfaces, and many but varied sites on the surfaces of amorphous precipitates of aluminosilicates, silica, or iron, aluminium, or manganese oxides at neutral or higher pH, or on the zone of alteration on the surfaces of crystalline or microcrystalline precipitates of oxides, carbonates, phosphates, etc. The retention of specifically adsorbed cations may become stronger with time as the cations diffuse into the body of a precipitate, or as local rearrangements of ions in the lattice increase their crystallinity. Thus McLaren et al. (1986) reported that the cobalt taken up by soil material from solution was mainly held on soil-derived oxide materials. In this form the cobalt was not readily desorbed and rapidly ceased to be isotopically exchangeable; in contrast, cobalt adsorbed onto soil organic matter remained isotopically exchangeable. In some cases crystallization of an amorphous precipitate may "squeeze out" the foreign cations it had occluded; they may then become concentrated in its outer, and still amorphous, zone (McLaren and Crawford, 1973; McBride, 1981).

4. *Held on insoluble organic materials*: this category includes trace metal cations in, or immobilized on, living or recently dead cells, but it consists mainly of cations complexed or chelated by organic materials, either recently synthesized, or the resistant residues from microbial metabolism. It may be convenient to distinguish "new" and "old" organic materials. The category includes both specific and nonspecific sites. Some of the organic matter of this group will be intrinsically insoluble; some of it will have been flocculated or precipitated by complexing cations, notably iron or aluminium, but also by basic cations like calcium or trace metals such as copper. Indeed, organic colloids may be closely associated with iron or aluminium oxides (Smith and Mitchell, 1984). Jeanroy et al. (1986b) referred to ferric organic complexes, more and less polymerized. It may be possible to disperse organic matter flocculated in this way by replacing its complexing cations with sodium or potas-

sium ions; much of the colloid material thus dispersed will not pass a dialysis membrane (Hodgson, 1963). The dispersed organic matter may still carry other, strongly adsorbed trace metal cations into solution or suspension with it.

5. *Held on, or occluded in, oxides of iron, aluminium, or manganese*: These oxides are rarely pure: they usually contain cations from each other, and probably trace metal cations as well. Norrish and Taylor (1961) reported that soil geothite (FeOOH) contained 15–30 moles per cent (mol%) of AlOOH, which increased its solubility in dithionite and may also account for some of the Al reported as being dissolved from soil materials by dithionite. Trace metal cations are easily adsorbed onto such precipitates as long as the host oxide is amorphous. Initially the adsorbed trace metals will be held in exchangeable form, but these may pass into more covalent (specific) forms with time. In general the surfaces of hydrous iron and manganese oxides are strong "scavenging agents" for heavy metals (Chao, 1972). Their capacity to adsorb trace cations from solution is controlled by their pH and crystallinity. As the oxides crystallize, the adsorbed cations may become occluded in the crystal structures, or they may be expelled toward the still amorphous surface layers.

   Different authors have recognized varied degrees of crystallinity. Some distinguish "amorphous" or "x-ray amorphous" from "crystalline" or "x-ray crystalline." Others distinguish several stages over the range from well crystallized through poorly crystallized, to amorphous or "loose gel" forms; these show a corresponding range of solubility in dithionite (Pawluck, 1972) or other extractants. Similarly, pyrophosphate at pH 10 dissolves iron from freshly precipitated iron hydroxides, but not from even poorly crystallized iron oxides; at pH 7, pyrophosphate dissolves iron from moderately crystallized geothite as well. Bascomb (1968) distinguished "aged" amorphous hydrous oxides from "gel" amorphous hydrous oxides, and Jeanroy et al. (1986b) distinguished more and less polymerized ferric organic complexes. Amorphous and paracrystalline compounds have been referred to as hydrous ferric oxides or hydroxides (Towe and Bradley, 1967) or as ferrihydrite ($Fe_5(OH)_8{\cdot}4H_2O$). However, few of these terms are standard. The crystallinity of the inorganic forms may be linked to their crystallite size as determined by Mössbauer techniques (Jeanroy et al., 1986a). Hydroxides or hydrous oxides are said to be less crystalline or more soluble than oxides (Jeanroy et al., 1986b).

   Iron and manganese oxides have also been distinguished as "easily," "moderately," or "difficultly reducible" according to the reagents that dissolve them. The ease of release of occluded trace metal cations is supposed to depend on the solubility of the oxides that contain them. Note that the term "free" iron, as commonly used, includes all iron present in crystalline or amorphous compounds, except that in the crystal lattices of nonoxide minerals.

6. *New "precipitates"* (carbonates, sulfides, phosphates, hydroxides, etc., but often not specified): The precipitates may be newly formed compounds of the trace metals themselves, although this is more likely in the digester than in the soil. If so, they usually contain more than one trace metal, and often form

mixed crystals or mixtures of crystallites with the corresponding salts of major elements, usually calcium or iron. Minor amounts of trace cations may be incorporated in new precipitates of compounds of calcium, magnesium, iron, or aluminum as they form, or may be adsorbed on the amorphous surfaces of such precipitates, whence they may be exchangeable, or may pass over into more "specific" adsorption when the surface layers crystallize.

III. Survey of Extractants

The following survey presents the extractants most commonly used and lists the groups of compounds it is presumed that they extract. The more aggressive extractants such as dithionite, acid ammonium oxalate, hydroxylamine hydrochloride, etc., are used to destroy certain categories of sites, and particularly the oxides and hydroxides of iron, manganese, or aluminium to release the trace metals they may also contain. Thus, this survey also discusses the efficiency of extractants at mobilizing iron, manganese, and aluminium, with the implication that an extractant that dissolves a particular category of iron oxide, etc., also mobilizes any trace metals that it contains.

Reaction times and the ratio of extractant to sample are not discussed. It is assumed that the published ratios of extractant to sample (soil or sludge) are high enough that the concentration of the extractant is not substantially reduced during the reaction, nor is its pH substantially changed by interaction with $CaCO_3$ or soil, and that the reaction times proposed were long enough. Reaction times are usually quite short (0.5–2.5 h) or adjusted to fit to the working day (e.g., 6–7 h or overnight, 16 h). It is not always clear that the composition of an extractant was *not* changed by its interaction with a sample.

A. Organic Solvents

Luoma and Jenne (1976) used 70% ethanol to break down viable cells, and to dissolve compounds of low polarity, and several workers have used acetyl acetone as a complexant (below). Apart from these, few sequences of extractants have included an organic solvent. This is perhaps surprising because sewage sludge, and other industrial wastes, often contain oily or waxy components that may protect parts of them from aqueous extractants, and some of which are very resistant to microbial decay in sludge-treated soils. Strachan et al. (1983), for example, found that 5–50% of the combustible components of digested sludges were soluble in *n*-hexane, which also dissolved 80% of the materials dissolved in a range of organic solvents. These solvents may remove small quantities of organically bound trace metals: Bergman et al (1979) found that small quantities of Cd, Pb, and Zn were dissolved in the nonpolar organic solvents they used for density separation of digested sludge. Whether or not there are significant amounts of these components, a preliminary extraction in *n*-hexane, followed by ether *or* acetone, may increase the efficiency of water-soluble extractants. Some of the organic materials dispersed by pyrophosphate are soluble in ether.

Because oven- or air-drying may affect the forms of combination of trace metals and may give rise to hydrophobic organic coatings on inorganic surfaces, then, if samples must be dried, it may be preferable to dry them with organic solvents (R.D. Macnicol and P.H.T. Beckett, unpublished).

B. Water

Several workers have measured a *water-soluble* fraction of the trace metals by shaking sludge or soil with water (unspecified) or with distilled or deionized water (Viets, 1962; Gupta and Chen, 1975; Sims and Patrick, 1978; Iyengar et al., 1981; Miller and McFee, 1983; Miller et al., 1986a,b). This usually precedes the measurement of exchangeable or specifically adsorbed forms. Wilber and Hunter (1979) measured the quantities that were dissolved in the river water to which the materials had been or would be exposed. Sims and Patrick (1978) divided the water-soluble fraction into free ions and ions complexed by organic matter by extracting the former with cation- and anion-exchange resins.

Sposito et al. (1982) used very pure distilled and deionized water (three repeats of a 2-h extraction) as an extractant for nonspecifically "sorbed" or "adsorbed" ions, *after* removing exchangeable ions with 0.5 *M* KNO_3. Lund et al. (1985) demonstrated that this reagent successfully removed cadmium, nickel, and zinc recently adsorbed onto freshly prepared hydrous iron oxides, and cadmium, lead, and zinc from freshly prepared hydrous aluminium oxides, and moderately successfully removed cadmium and lead, or copper and nickel, respectively, from these two groups of oxides.

Bradford et al. (1975) analyzed "saturation extracts" of rewetted samples of sludges and sludge-treated soils as surrogates for their equilibrium solutions. However, Eaton et al. (1960) have shown that soil saturation extracts tend to be more dilute than the equilibrium solution extracted from the same soils by suction, and Adams et al. (1980) have shown that drying and rewetting is likely to alter the composition of the equilibrium solution. This is better extracted by centrifuging undried material through a membrane, or in an organic displacent immiscible with and denser than water (Campbell and Beckett, 1988). Gupta and Chen (1975) separated "interstitial water" by squeezing it through a .05-μm millipore filter.

C. Simple Salts

Many workers have estimated "exchangeable" (Stover et al., 1976; Silviera and Sommers, 1977; Shuman, 1979; Harrison et al., 1981; Emmerich et al., 1982; Miller and McFee, 1983; Hickey and Kittrick, 1984; McLaren et al., 1986; Jarvis, 1986, Neilsen et al., 1986), "neutral salt exchangeable" (Miller et al., 1986a), "water-soluble and exchangeable," or "nonspecifically adsorbed" (McLaren and Crawford, 1973; Iyengar et al., 1981) ions of trace metals from soil, sludge, etc. by extraction with relatively strong solutions of the chlorides or nitrates of potassium, magnesium, and calcium. For example:

KCl	2 *M* (Himes and Barber, 1957)
KNO_3	0.5 *M* (Emmerich et al., 1982; Sposito et al., 1982)); 1 *M* (Stover et al., 1976; Silviera and Sommers, 1977; Schalscha et al., 1982; Miller and McFee, 1983)
$CaCl_2$	0.05 *M* (McLaren and Crawford, 1973; Shuman, 1979; Iyengar et al., 1981); 0.5 *M* (McLaren et al., 1986)
$Ca(NO_3)_2$	0.25 *M* (Miller et al., 1986b)
	0.5 *M* (Tiller et al., 1972); Miller et al., 1986a,b)
$MgCl_2$	0.5 *M* (Gibbs, 1973, 1977)
	1 *M* (Schuman, 1979; Tessier et al., 1979; Harrison et al., 1981; Hickey and Kittrick, 1984; Neilsen et al., 1986)
$LaCl_3$	0.3 *M* (Jarvis, 1986).

It is presumed that these reagents act as sources of cations to displace trace metals held on inorganic and/or organic sites predominantly by electrovalent forces.

Neutral salts like these have the advantage that they will barely affect the operative pH at the exchange sites, however much their effect on the diffuse layer around such sites may alter the measured pH of the equilibrium solution. They have been preferred to ammonium salts, and particularly to ammonium acetate, on the grounds that ammonium and acetate ions may have specific complexing effects as well (Neilsen et al., 1986). Thus, ammonium chloride and acetate have been supposed to dissolve transition metals from coatings on sediment grains, whereas 0.5 *M* $MgCl_2$ does not (Gibbs, 1973, 1977). Also, ammonium acetate mobilizes more organic matter than does 0.05 *M* $CaCl_2$ (McLaren and Crawford, 1973).

In general, the cations that are the harder Lewis acids are likely to be better displacents of the metal cations that are held by more covalent forces; for example, magnesium is a stronger displacent than calcium in displacing cobalt and zinc from minerals (Tiller et al., 1972). Magnesium salts have been used to estimate "specifically adsorbed" (Tessier et al., 1979; Iyengar et al., 1981) or "adsorbed" metals (Gibbs, 1973, 1977) and more strongly held forms in general (Tiller et al., 1972). Jarvis (1986) found that 0.3 *M* $LaCl_3$ extracted substantially more exchangeable Al from soils than 1 *M* KCl, some of it probably from clay interlayers and organic complexes; $LaCl_3$–Al was strongly correlated with EDTA-Al.

As between the nitrate, chloride, and sulfate of magnesium, the last two extracted more zinc from soils than the nitrate, which Tiller et al. (1972) attributed to the greater stability of their anion complexes with zinc.

D. Less Simple Salts

This poorly defined group comprises the salts of soft or borderline cations, such as copper(II) or lead(II), able to displace similar cations from sites (organic or inorganic) where they are retained by partially covalent or coordinate forces.

Miller et al. (1986b) used 0.05 *M* lead nitrate to displace "specifically bound" copper and 0.05 *M* lead nitrate plus 0.1 *M* calcium nitrate to measure "Pb-displaceable" metals (Miller et al., 1986a), which they supposed were metal ions bound covalently to oxides or organic groups. Under other circumstances lead nitrate has been no more effective than calcium nitrate. Lead nitrate extracts less trace metals than 0.44 *M* acetic acid, so the latter has been assumed to dissolve some poorly crystalline hydroxy- and carbonate phases as well. The total quantity of trace metals displaced by acetic acid and lead nitrate has been reported to be more or less constant, regardless of which was used first (Miller et al., 1986a).

Copper acetate (0.125 *M*) has been used to displace "complexed metals on organic matter or hydrous oxides" (Soon and Bates, 1982) where the quantity displaced was significantly correlated with the organic matter present, and to release organically complexed zinc (Sedberry and Reddy, 1976); Murthy (1982) and Mandal and Mandal (1987) used 0.05 *M* copper acetate, in the latter case to measure "organic-complexed" metals. Copper acetate displaces considerably more zinc from organic combination than 2 *M* KCl (Himes and Barber, 1957). Jarvis (1986) has suggested that copper(II) might be used to displace aluminium complexed with organic matter; he found that copper(II) was as efficient as lanthanum.

E. Acetic Acid

Acetic acid, alone or buffered by acetate salts, has been used for a long time to measure the "availability" of plant nutrients in the soil. Its employment in studies on trace metals was probably inevitable. It may have been a mistake, except perhaps that in reactions that neutralize hydrogen ions a weak acid will hold its pH better than a strong acid.

Molar acetic acid has been assumed to dissolve carbonates and some iron and manganese oxides (Gupta and Chen, 1975). A mixture of 0.44 *M* acetic acid plus 0.1 *M* calcium acetate has been used to dissolve "acid-soluble" trace metals (Miller et al., 1986a). Solutions of 2.5% acetic acid (0.42 *M*; pH 2.5) or 0.44 *M* or 0.5 *M* acetic acid have been reported to extract "specifically bound" or "specifically adsorbed" trace metals (Iyengar et al., 1981; Miller et al., 1986b; Rappaport et al., 1986), supposedly from sites associated with oxides (McLaren and Crawford, 1973). The reagent is said to be less specific than lead nitrate. It may dissolve poorly crystalline hydroxy- and carbonate phases as well as displacing covalently bonded trace metals (Miller et al., 1986b): the amount of soil cobalt extracted by 0.5 *M* acetic acid was correlated with the total iron contents of the same soils, but not with their total manganese (Jarvis, 1986; McLaren et al., 1986). Luoma and Jenne (1976) reported that acetic acid extracted cadmium, cobalt, and zinc recently adsorbed onto iron oxides, decomposed organic matter, carbonates, or manganese oxides. Stover et al. (1976) reported that 2.5% acetic acid dissolved zinc and copper carbonates, but not cadmium carbonate and only a little lead carbonate or zinc and copper sulphides. Rendell et al. (1980) extracted sediment cores with 25% (4.2 *M*) acetic acid (pH 1.8) spiked with copper, and found that 31% of the copper originally present in the solution was adsorbed overnight.

F. Acetates

Solutions of ammonium acetate and ammonium chloride (Gibbs, 1973, 1977; Shuman, 1979) have been widely used to measure soil cation-exchange capacity (CEC), or exchangeable ions, because (i) they are well buffered; (ii) the ammonium ion blocks the slow release of interlamellar cations from the weathered periphery of clay lamellae (Scott and Welch, 1961) and gives a sharp CEC; and (iii) in the early days, because excess ammonium acetate could be disposed of relatively easily by oxidation or ignition. Thus, their use has spread to estimates of "exchangeable" trace metal ions in other materials; 1 *M* ammonium acetate at pH 2.5 has been used to measure "ion-exchangeable" trace metals (Förstner et al., 1981), and, at pH 6.7 or pH 7, to measure "exchangeable," "exchangeable non-specific" or "ion-exchangeable" trace metals (Viets, 1962; Luoma and Jenne, 1976; Wilber and Hunter, 1979; Rendell et al., 1980; Förstner et al., 1981; Iyengar et al., 1981; Soon and Bates, 1982); 5 *M* ammonium chloride at pH 8 has been used to measure "exchangeable" cations (Shuman, 1979), and 0.5 *M* calcium acetate to measure "exchangeable" copper (Miller et al., 1986a). It has been supposed that ammonium acetate removes chromium and zinc from sites on organic matter and iron oxides (Luoma and Jenne, 1976), that it can dissolve oxide coatings in the hydrous oxide fraction (Shuman, 1979), and that ammonium chloride or acetate can dissolve or complex transition metals from coatings on sediment grains (Gibbs, 1973, 1977). Ammonium acetate has also been added to other, stronger extractants to block the readsorption of the trace metals they have caused to be released (Förstner et al., 1981; Hickey and Kittrick, 1984).

1 *M* sodium acetate has been used at soil pH to measure exchangeable ions (Sims and Patrick, 1978); at pH 5 to dissolve carbonates or "metals bound to carbonate minerals" (Förstner et al., 1981; Harrison et al., 1981); and at pH 8.2 to measure exchangeable ions *without* dissolving carbonates (Tessier et al., 1979). It has also been reported to extract metals bound to or occluded in iron oxides (Hickey and Kittrick, 1984), and possibly to attack iron or manganese oxides, but to have no effect on organic matter or free iron oxides (Tessier et al., 1979). It is assumed (Luoma and Jenne, 1976; Tessier et al., 1979) that the acetate anion has some complexing effects.

G. Oxalic Acid and Oxalates

Chao and Zhou (1983) used 3% oxalic acid (0.33 *M*) at 92°C to dissolve free oxides and primary minerals by complexation, and Ball and Beaumont (1972) used it in place of the more commonly used acid ammonium oxalate. Le Riche and Weir (1963) reported that oxalic acid attacked silicate minerals. Sodium oxalate was reported by Olmstead et al. (1930) to be a very strong complexant for calcium, and therefore a better soil dispersant than NaOH or Na_2CO_3.

H. Sodium Citrate

Solutions of trisodium citrate (0.25 *M*) have been used as extractants for iron oxides (Stover et al., 1976) or trace metals in iron oxides. Rendell et al. (1980) used 0.1 *M* trisodium citrate to extract sediment cores, but found that 30% of the copper originally in solution was readsorbed. Citrate is commonly added to the dithionite extractant (p. 163) to facilitate the dissolution of iron oxides by complexing the iron as it is released. However, Saunders (1959) showed that citrate alone complexed 70–90% as much as Al, and 20–90% as much Fe, as were extracted by both reagents together. Citrate has also been used to dissolve aluminium coatings (Mehra and Jackson, 1960) in order to release associated silica cements.

I. Fluorides

Potassium and sodium fluorides have been used as extractants for trace metals in soils and sludges mainly because of the complexing power of the fluoride anion. Ammonium fluoride (0.5 *M*; pH 6.5) has been used as a complexant for aluminium compounds, but Stover et al. (1976) argued against it on the grounds that there will also be complexation by the ammonium cations and the results may be difficult to interpret. Schalscha et al. (1982) used 0.5 *M* NaF (pH 6.5) to extract exchangeable trace metals from sludge-treated soils. Schalscha et al. (1980) reported that 0.5 *M* NaF mobilized less aluminium than 0.5 *M* KF. Stover et al. (1976) recommended 0.5 *M* KF (pH 6.5) as an extractant for "adsorbed" metals, confirming that it does not dissolve carbonates or sulfides and should not mobilize trace metals from organic matter: for this reason it comes *before* pyrophosphate in their sequence of extractants. Potassium fluoride (0.5 *M*) displaces much less zinc than 1% acetic acid.

On the other hand, 0.5 *M* NaF has also been reported (Bremner and Lees, 1949) to mobilize soil organic matter by precipitating calcium fluoride and complexing other metals. It raised the pH considerably, but extracted less organic matter than 0.1 *M* sodium pyrophosphate. Schnitzer and Wright (1956) reported that an HCl-HF mixture used to remove the mineral components of a soil also mobilized soil organic matter flocculated by aluminium or iron complexation.

The most specific effect of fluoride extractants appears to be on amorphous ("gel-like" or "poorly ordered") aluminosilicate gels and allophane (Bracewell et al., 1970; Loveland and Bullock, 1976), where it extracts aluminium that would also be extracted by acid ammonium oxalate (dark) or dithionite extractants. The resulting release of hydroxyl is a well established test for allophane and amorphous aluminium: Loveland and Bullock (1976) reported that the rate of hydroxyl production is a measure of the crystallinity of the amorphous aluminosilicates being dissolved, the rate being the highest for the least crystalline. Rich and Obenshain (1955) used 1 *M* NH_4F to clean polymerized $Al(OH)_2^+$ coatings from clays.

J. Acetyl Acetone

To date acetyl acetone has been mainly used to complex and remove iron and aluminium from soil organic matter, in order to disperse it or to leave it in a form more easily mobilized by other means (Hamblin and Posner, 1979). Thus Scott and Anderson (1976) extracted soils with 0.2 *M* aqueous acetyl acetone at pH 8.0 after they had been dispersed by ultrasonication. This mobilized quantities of iron and aluminium comparable to those extracted by Tamm's (1922) acid oxalate reagent, but substantial parts of the mobilized iron and aluminium remained bound to dispersed organic matter.

Martin and Reeve (1957) used 0.2 *M* aqueous acetyl acetone to remove iron and aluminium as acetyl acetone complexes, in order to disperse organic matter. The amounts of carbon or aluminium thus mobilized increased with the pH of the extractant, although the amount of iron decreased. The dispersed organic matter and the iron and aluminium complexed in it were extractable in ether. Hamblin and Posner (1979) reported that aluminium and iron complexes with acetyl acetone were very stable in nonpolar solvents. These authors supposed that acetyl acetone dispersed less aluminium or carbon than sodium pyrophosphate because the latter provides abundant sodium ions to replace the complexed aluminium, whereas acetyl acetone could provide H_3O^+ only by ionization. Acetyl acetone partially dissolves free iron and aluminium hydroxides.

Hamblin and Posner (1979) referred to the use of acetyl acetone in benzene to remove iron and aluminium from soils. They used 5% solutions (0.5 *M*) of acetyl acetone in deionized water or in toluene, adding a very little water to the latter to promote ionization. Both solutions were able to mobilize aluminium and iron from their hydroxy-metal polymers, and relatively more from the less crystalline forms of these. The aqueous solution removed relatively more iron and aluminium from organic materials, and the toluene solution more from inorganic materials.

K. EDTA

EDTA has been used in various forms, not always very precisely recorded, but mainly at pH 6.5–7 or at pH 10, being titrated to the required pH with NaOH or ammonia. The natural pH values of 0.1 *M* disodium or diammonium EDTA are pH 4.5–4.8; and pH 10 for 0.1 *M* tetrasodium-EDTA.

The following roles have been claimed for EDTA.

1. *To dissolve carbonates*:

0.1 *M* EDTA (pH 6.5) (following pyrophosphate and fluoride) as an extractant for carbonates, was shown to dissolve cadmium, copper, lead, and zinc carbonates, and partially to dissolve lead sulfides but not other sulfides (Stover et al., 1976);

0.05 *M* Na_2-EDTA (after NaOH extractants for organically bound elements) to dissolve carbonates but possibly not sulfides (Emmerich et al., 1982; Sposito et al., 1982);

0.1 *M* EDTA (pH 6.5) (following 1 *M* KNO_3 and pyrophosphate and fluoride) to extract "inorganic precipitates" (Rendell et al., 1980);

0.1 *M* EDTA (pH 7.0) to dissolve carbonates by complexation and thereby mobilize the copper occluded in them (Miller et al., 1986b);

0.5 *M* Na_2-EDTA substantially dissolved pure samples of cadmium, copper, nickel, lead, and zinc carbonates, and some nickel and lead sulfides, but not other sulfides (Lund et al., 1985).

2. *To extract trace metals from the oxides or hydroxides of iron*, or to dissolve them and thereby to release trace metals adsorbed on or occluded in them:

0.1 *M* EDTA + 1 *M* NaOH to remove cadmium, cobalt, and zinc freshly adsorbed onto iron oxides (Luoma and Jenne, 1976);

0.1 *M* EDTA to extract trace metals occluded in noncrystalline or poorly crystalline iron oxides (Miller and McFee, 1983);

0.1 *M* EDTA to extract some oxides (but less rigorously than pyrophosphates) (McLaren and Crawford, 1973);

0.4 *M* Na_2-EDTA extracted cobalt from soil, and specifically from oxide surfaces, in quantities proportional to the "total" soil iron and "total" soil manganese: it extracted more cobalt than was isotopically exchangeable, and more cobalt than pyrophosphate or hydroxylamine (McLaren et al., 1986);

EDTA (pH 10) was used to remove amorphous iron oxides (Jeanroy et al., 1986a);

0.1 *M* Na_2-EDTA (pH 4.4–6.0, buffered by ammonium acetate/acetic acid) was believed to extract x-ray amorphous iron oxides (Borggard, 1976);

Na_4-EDTA dissolved amorphous iron compounds, but not crystalline iron forms (Jeanroy et al., 1986b);

0.1 *M* EDTA (pH 10, 80°C) dissolved iron complexed by organic matter and also amorphous iron oxides, as pyrophosphate does (Jeanroy and Guillet, 1981).

Le Riche and Weir (1963) reported that ultraviolet radiation increased the efficacy of the EDTA extractant.

3. *To release organically bound trace metals*:

0.4 *M* EDTA is usually thought to be an organic extractant (McLaren et al., 1986);

Na_4-EDTA does not extract iron from organic chelates (Jeanroy et al., 1986b);

NaOH/EDTA extracts organically bound iron (Cottenie et al., 1979);

0.4 *M* Na_2 EDTA reportedly dissolves cobalt chelated on organic matter (McLaren et al., 1986);

0.1 *M* EDTA (pH 7.0) extracts copper from organic ligands (Miller et al., 1986b);

0.1 *M* EDTA + 1 *M* NaOH removed cadmium, cobalt, and zinc freshly adsorbed onto organic matter (Luoma and Jenne, 1976).

McLaren and Crawford (1973) showed that 0.1 *M* EDTA was less effective than pyrophosphate at dispersing organic matter.

In addition, ammonium EDTA (pH 7.0) has been used to extract exchangeable and other inorganic forms of aluminium (Jarvis, 1986), although not from aluminosilicate minerals, and to remove a general category of adsorbed, chelated, or complexed trace metals (Viets, 1962). Jarvis (1984) used 1% HNaCa-EDTA, in ammonium acetate at pH 8.0, to extract manganese.

L. DTPA

DTPA has been used to extract exchangeable and organically bonded trace metals, and to dissolve "precipitates" (Schalscha et al., 1982). More generally, 0.005 *M* DTPA + 0.01 *M* $CaCl_2$ + 0.1 *M* triethanolamine (pH 7.3) has been used (Silviera and Sommers, 1977) to measure "available" forms. Hickey and Kittrick (1984) showed that DTPA-Zn is correlated with pyrophosphate-Zn and assumed that DTPA extracts organically bound zinc.

Brown et al. (1971) showed that the quantities of zinc extracted from a range of soils by DTPA, EDTA, or dithizone (diphenyl thiocarbazone) were mutually correlated.

M. Sodium/Potassium Pyrophosphate

Pyrophosphate was originally introduced into studies on soil flocculation and soil structure as an extractant for soil organic matter, and particularly for the organic matter flocculated and complexed by calcium, or by iron and aluminium, which it dispersed by complexing the flocculating cations.

Bremner and Lees (1949) introduced the use of pyrophosphate as a dispersing agent for soil organic matter on the grounds that the efficiency of a neutral salt at dispersing organic matter depends on its efficiency at removing metal ions from organic complexes by forming soluble complexes with them. Neutral pyrophosphate proved to be the most efficient of the neutral reagents they examined. McLaren and Crawford (1973) reported that it extracted more colloidal organic matter than EDTA, and Miller et al. (1986a) reported that it mobilized more organic matter than H_2O_2 or NaOCl, and that its efficiency increased with pH. Bremner and Lees (1949) recommended pH 8 as a compromise between complexation and the precipitation of hydroxides. They also reported that the organic matter extracted did not pass a dialysis membrane; once it had been freed of soluble salts by dialysis it could be reflocculated by calcium, copper, iron, or manganese, for example, and, if required, then be dissolved again in pyrophosphate. Many workers have shown that pyrophosphate can dissolve some fractions of iron, aluminium, and manganese oxides, presumably with any trace metals they contain.

It is not clear whether the workers who first used pyrophosphate as an extractant for "organically bound" trace metals in soil or sludge, then believed that it worked by (1) removing trace metals from ligand sites on organic material by forming stronger soluble complexes, or (2) mobilizing those parts of the organic matter (with the trace metals still complexed on them) that were flocculated by iron, aluminium, or calcium, by complexing these cations.

If the latter is the case, then it may be a mistake to clarify the extract by centrifugation, or by adding a flocculant, because reprecipitating the dispersed organic matter also removes from the extract any trace metals it still holds.

Much of the early work used potassium pyrophosphate (e.g., Bascomb, 1968), but potassium or sodium pyrophosphate are now used more or less indiscriminantly: Loveland and Digby (1984) reported that their effects are very similar. Either because it mobilizes iron and aluminium as well as organic matter, or because much soil organic matter is relatively inert, Schalscha et al. (1982) found that the quantities of trace metals extracted from a soil by pyrophosphate were not correlated with its total organic matter content. Jarvis (1984), on the other hand, found that the considerable quantities of manganese it extracted from soils were in proportion to their total organic carbon. Rendell et al. (1980) claimed that pyrophosphate dissolves "free" oxide associated with organic colloids, particularly of iron and aluminium, but much less "free" iron from inorganic materials.

McKeague (1967) reported that "only iron or aluminium associated with organic matter" are mobilized by 0.1 *M* sodium pyrophosphates, although less completely than by acid ammonium oxalate, and that pyrophosphate did not extract iron from amorphous or crystalline oxides; Miller et al. (1986a) suggested that pyrophosphates have little effect on most iron oxides; Jeanroy et al. (1986b) reported that pyrophosphate dissolves less iron than acid ammonium oxalate: they assumed that it did not dissolve inorganic iron compounds however amorphous, but that it did dissolve "hydroxy-ferric-organic complexes," more strongly than did tetraborate, which only dissolves their less polymerized forms. Rappaport et al. (1986) reported that pyrophosphate probably extracts some noncrystalline oxides and hydroxides of aluminium, iron, and manganese; McLaren and Crawford (1973) suggested that it dissolves some oxides but less than does EDTA.

Bascomb (1968) reported that 0.1 *M* potassium pyrophosphate *at pH 10* extracted iron from "freshly precipitated hydrous ferric hydroxide" (later called "amorphous gel"), but very little from "partly crystallized," "x-ray amorphous," or "moderately crystallized oxides," and that it also peptized clay-sized mineral matter. It dissolved all Fe–organic complexes but hardly attacked silicates. It may be able to dissolve very finely divided material, whatever its crystallinity. *At pH 7* it appears to extract all organic–Fe complexes, and amorphous "aged" as well as amorphous "gel" precipitates, and some iron from moderately crystallized goethite, and from all but well crystallized oxides. Bascomb offered the two reagents (pyrophosphate at pH 10 or pH 7) as a means for distinguishing the two forms of amorphous precipitate.

Various workers have used 0.1 *M* or 1 *M* sodium or potassium pyrophosphate, almost always at or near room temperature. In most cases the extractant is used at its natural pH, which is pH 10 for 0.1 *M* potassium pyrophosphate, but often this is not specified. Pyrophosphate has been used to extract trace metals that are "organically bound," or held by "specific adsorption" on organic sites (McKeague, 1967; Bascomb, 1968; McLaren and Crawford, 1973; Sims and Patrick, 1978; Iyengar et al., 1981; Schalscha et al., 1982; Miller and McFee, 1983; Jarvis, 1986; Jeanroy et al., 1986b; McLaren et al., 1986; Miller et al., 1986a,b; Rappaport et al., 1986). Iyengar et al. (1981) found that the pyrophosphate Zn in a soil was directly correlated with its total organic carbon and "free" manganese, and inversely with its "free" iron.

The picture is complicated, however, by the fact that pyrophosphate mobilizes organic matter by peptizing organic complexes rather than by dissolving them, and amorphous colloids or gels of metal organic complexes are dispersed at the same time (Hodgson, 1963; Stover et al., 1976; Loveland and Bullock, 1976). Thus Bascomb (1968) reported that as much as half of the iron extracted in 0.1 *M* potassium pyrophosphate at pH 10 was organically complexed iron, unable to pass a 10-μm millipore filter, and flocculated by 0.880 ammonia. He distinguished forms that were precipitated in this way as organically bound, and forms that were not as inorganically (pyrophosphate) bound. Loveland and Digby ((1984) showed that some of the iron (and aluminium) in pyrophosphate extracts was colloidal, and could be removed by centrifugation or flocculants: McKeague (1967) found that it required strong centrifugation (20,000 × *g*) to clear the extract, and that the concentrations of iron and aluminium in the supernatant were much less than in the original suspension.

Jeanroy and Guillet (1981) showed that 0.1 *M* potassium pyrophosphate at pH 10 dissolved some iron complexed by organic matter, but also dispersed (not dissolved) finely divided ferruginous precipitates, and notably the colloidal aggregates of small ferruginous spheroids that under some circumstances are formed in the soil by the precipitation of mobile organic–Fe complexes. They reported that pyrophosphate did not *dissolve* very much iron even from poorly crystalline oxides, and that EDTA and pyrophosphate *dissolved* similar quantities of iron, but EDTA did not *peptize* ferruginous colloids as well.

In addition, McLaren et al. (1986) and Miller et al. (1986a) reported that 0.1 *M* sodium or potassium pyrophosphates dissolved reducible forms of manganese compounds and they partially dissolved Mn oxides. Over 24 h, pyrophosphate dissolved most of the copper held in manganese oxides, but had little effect on most iron oxides. Martin and Reeve (1957) suggested that sodium pyrophosphate was a more efficient extractant for aluminium (particularly) and iron than acetyl acetone because it carried sodium ions to replace them. At pH 7 it dissolved substantial amounts of soil aluminium. Jarvis (1986) found that the exchangeable aluminium dissolved from a soil by pyrophosphate was correlated with its total organic matter. Coffin (1963) reported that 0.1 *M* pyrophosphate dissolved lead carbonate, but little of zinc or copper carbonates or sulfides. Lester et al. (1983) reported that it was a good solvent for lead carbonate, and Miller et al. (1986b) reported that it dissolved some malachite and copper sulfides, but not copper oxide. It has been reported to extract more aluminium (Jarvis, 1986) and less cobalt (McLaren et al., 1986) than EDTA.

N. Acid Ammonium Oxalate (and Ammonium Tartrate)

Acid ammonium oxalate extractants are direct descendants of Tamm's (1922) acid oxalate reagent. In dissolving iron and aluminium oxides, etc., they release the trace metals adsorbed on or occluded in them. Usually the ammonium oxalate-oxalic acid mixture is used at pH 3.25 (sometimes quoted as pH 3), corresponding to 0.175 *M* ammonium oxalate + 0.1 *M* oxalic acid, but other proportions have been used (such as 0.17 *M* ammonium oxalate and 0.15 *M* oxalic acid; McLaren et al., 1986). McKeague and Day (1966) used pH 3 as a compromise between pH 2, at which silicate minerals may be attacked, and pH 4, which is less well buffered.

The crucial point is that the action of acid ammonium oxalate is very sensitive to illumination. As early as 1954, Mitchell and Mackenzie had suggested that the reagent might be more effective in sunlight. Later, de Endredy (1963) demonstrated that its potency was very much increased by illumination in the near ultraviolet (centered on 366 nm), where it dissolved even haematite, and bleached soil samples by freeing them of all iron oxides. Schwertmann (1964) used 0.2 *M* ammonium oxalate + 0.2 *M* oxalic acid at pH 3 and showed that its potency increased in this order: dark < diffuse illumination < sunlight < ultra-violet. Under ultraviolet light, it was as effective as dithionite.

Most workers have adopted the two extremes, that is, in the dark or in ultraviolet illumination, the former being supposed to dissolve amorphous iron oxides and the latter to also dissolve crystalline oxides (e.g., Schwertmann, 1964; McLaren and Crawford, 1973; Stover et al., 1976; Sims and Patrick, 1978; Shuman, 1979; Förstner et al., 1981; Iyengar et al., 1981; Jeanroy et al., 1986a,b; McLaren et al., 1986; Miller et al., 1986,a,b; Rappaport et al., 1986). Some workers use slightly different solutions in the dark and under ultraviolet (e.g., Jarvis, 1986): Under ultraviolet, the reagent is commonly heated to 80–100°C.

However, the antithesis between the reactions in the dark and under ultraviolet is not quite so sharp as might be implied. Thus acid ammonium oxalate *in the dark* has been shown to dissolve:

Magnetite partially, but very little crystalline goethite or bentonite (Chao and Zhou, 1983);

X-ray amorphous iron oxides or hydroxides (Schwertmann, 1964; Smith and Mitchell, 1984);

Amorphous "aged" and "gel" hydrous oxides, and fulvic but not humic iron complexes, and in this respect to be equivalent to pyrophosphate at pH 7 (Jeanroy et al., 1986b);

Poorly crystallized iron oxyhydroxides (Förstner et al., 1981);

Only those "x-ray-amorphous" iron oxides that are the "amorphous products of recent weathering"; it dissolved little iron from goethite or haematite, but was more effective than dithionite at dissolving amorphous aluminium (McKeague and Day, 1966);

A little iron from crystalline oxides, but significant amounts from amorphous iron oxides, and to dissolve more iron and aluminium than pyrophosphate (McKeague, 1967);
To have little effect on crystalline minerals (Smith and Mitchell, 1984);
To take a very long time to bleach a sample free of all iron colors (Mitchell and Mackenzie, 1954).

However, although it dissolves no more of the oxides than EDTA in the first 4–5 h, acid ammonium oxalate will dissolve all iron oxides in time, unlike EDTA (Borggard, 1976).

Under ultraviolet illumination, the same or similar reagents dissolve:

All iron forms, including Fe–organic complexes (Jeanroy et al., 1986b);
All iron oxides (Miller et al., 1986a);
Goethite and other iron oxides, but not gibbsite or boehmite (probably the aluminium extracted came from inclusions in the goethite lattice), to leave a grey bleached residue, which was washed with fresh oxalate solution, and then with 0.1 *M* KCl, to recover readsorbed iron. It also extracted some organic matter, probably with bound trace metals (Le Riche and Weir, 1963);
Crystalline and noncrystalline aluminium hydroxides and oxyhydroxides (Jarvis, 1986);
More iron than conventional dithionite, even from haematite, but did not remove iron from a ferriferous montmorillonite: it left the sample wholly bleached. Under ultraviolet radiation the pH seemed not to matter as long as pH > 7 (de Endredy, 1963).

Numerous workers have used acid ammonium oxalate in the dark or with ultraviolet radiation to dissolve trace metals occluded in amorphous or crystalline iron oxides (McClaren and Crawford, 1973; Miller et al., 1986a,b). Iyengar et al. (1981) found that the zinc extracted from a soil was proportional to its "free" iron and "free" aluminium contents. Jarvis (1984) used acid ammonium oxalate at pH 3.25 (in ultraviolet) to extract manganese from soils, and found that it removed considerably more than did a 0.4% solution of dithionite in 1 *M* ammonium acetate at pH 7.

As well as iron oxides, acid ammonium oxalate appears to extract iron or aluminium from organic complexes, possibly more efficiently than pyrophosphate (Le Riche and Weir, 1963; McKeague, 1967; Chao and Zhou, 1983; Jarvis, 1986; Jeanroy et al., 1986b), and it disperses significant amounts of organic matter (Le Riche and Weir, 1963; Kuo et al., 1983), which may carry complexed trace metals. It also dissolves zinc and copper carbonates, but not lead carbonate; it may or may not attack sulfides (Stover et al., 1976; Miller et al., 1986b). Wang and Schuppli (1986) used acid ammonium oxalate to estimate extractable silica in soils. The low pH may damage clay minerals (Mitchell and Mackenzie, 1954; Le Riche and Weir, 1963).

Dion (1944) also explored the use of ammonium tartrate and aluminium at pH 6.4 as a reducing agent to dissolve iron oxides; the reaction mixture was kept gently boiling. The reagent attacked nontronite, but no other clays. Mitchell and MacKenzie (1954) found it relatively inefficient.

O. Hydroxylamine Hydrochloride (and Hydroquinone)

Hydroxylamine hydrochloride ($NH_2OH{\cdot}HCl$, or HH) at pH 1 or 2 is a mild reducing agent commonly used to mobilize manganese from its oxides, or trace metals adsorbed on or occluded in manganese oxides. Its efficiency increases with temperature and acidity. Unacidified 0.1 *M* hydroxylamine hydrochloride has a pH of 3.6, so it is usually acidified with dilute nitric, hydrochloric, or acetic acid.

Thus, Chao (1972) reported that 0.1 *M* HH in 0.1 *M* HNO_3 (pH 1) rapidly dissolved most of the manganese in several soils, but only a few per cent of soil iron, while 0.1 *M* HH in 0.01 *M* HNO_3 (pH 2) dissolved somewhat less manganese and very little iron. He claimed to find a sharp cutoff between fractions that were soluble or not soluble at the higher pH. However, he recommended pH 2 to discriminate between the trace metals in iron or in manganese oxides, and to avoid hydrolysis of the cations released. Chao and Zhou (1983) reported that 0.25 *M* HH in 0.25 *M* HCl at 70°C dissolved substantial amounts of amorphous iron oxides and significant amounts of crystalline iron oxides: at 50°C, it dissolved very little iron from crystalline oxides but still dissolved amorphous iron oxides, to the same extent as acid ammonium oxalate in the dark. Miller et al. (1986b) found that 0.01 *M* HH in 0.1 *M* HNO_3 dissolved reducible soil manganese but not reducible iron compounds, and had no significant effect on most iron oxides, although 0.25 *M* HH in 0.25 *M* HCl dissolved noncrystalline iron oxides.

Thus, 0.1 *M* HH in 0.2 *M* HNO_3 has been used to dissolve manganese concretions (Stover et al., 1976), trace metals occluded in manganese oxides (Miller and McFee, 1983; Miller et al., 1986a) or "easily reducible manganese oxides" (Wilber and Hunter, 1979); 0.1 *M* HH in 0.01 *M* HNO_3 at pH 2 has been used to dissolve trace metals occluded in manganese oxides (Miller et al., 1986b; Neilsen et al., 1986) and easily reducible manganese, from amorphous or partly amorphous iron oxides (Gupta and Chen, 1975; Sims and Patrick, 1978; Förstner et al., 1981; McLaren et al., 1986). Unacidified 0.1 *M* HH at pH 3.6 has been reported to dissolve birnessite but not lithiophorite or goethite: poorly crystalline forms are more soluble, and iron and aluminium oxides only weakly soluble. At lower pH it may dissolve other oxides (Tokashiki et al., 1986). Iyengar et al. (1981) found that the zinc it extracted from a soil was correlated with the free manganese in the soil and with soil pH. Tessier et al. (1979) found that it extracted less than EDTA, and Luoma and Jenne (1976) found that it displaced freshly adsorbed trace metals from carbonates, oxides, and organic matter. Hydroxylamine hydrochloride has been reported to dissolve zinc and other carbonates, but not lead carbonates or sulfides (Stover et al., 1976; Förstner et al., 1981).

A solution of 0.25 *M* HH in 25% acetic acid at 70°C had only limited effects on crystalline iron oxides, and extracted much the same amorphous oxides as acid ammonium oxalate (Chao and Zhou, 1983); at 96–100°C, 0.04 *M* HH in 25% acetic acid (pH 2) extracted trace metals bound to iron/manganese oxides of a range of crystallinities, and a significant fraction of any aluminium oxides (Harrison et al., 1981; Hickey and Kittrick, 1984) or "moderately reducible iron oxides" present (Gupta and Chen, 1975). It did not dissolve sulfides or organic matter (Tessier et al., 1979).

Hydroxylamine has the advantage that it does not add new compounds to the reaction mixture, unlike the sulfides precipitated by dithionite (Tessier et al., 1979), but there is substantial readsorption of displaced cations from the reaction mixture (Rendell et al., 1980; Förstner et al., 1981).

Jarvis (1984) has explored the use of 0.2% hydroquinone in 1 *M* ammonium acetate at pH 7 as a mild reducing agent to extract easily reducible manganese: it dissolved considerably less manganese than hydroxylamine hydrochloride.

P. Sodium Dithionite

Sodium dithionite is used as an extractant because of its capacity for reducing even crystalline iron oxides and releasing occluded trace metals. A 4% solution of sodium dithionite is about pH 3.5, and dithionite at pH 3.5–3.8 has been reported to dissolve all iron oxides and most organic–iron complexes (Bascomb, 1968). In reactions at pH < 4.5 it precipitates sulfur and iron(II) sulfides (Coffin, 1963) so it is usually used at pH ≥ 4.75 and often between pH 5.8 and 7.3. Mehra and Jackson (1960) suggested that pH 7.8 is optimal. Mitchell and Mackenzie (1954) washed the reaction mixture repeatedly with dilute HCl to remove iron(II) sulfides deposited on the active surfaces. They removed any precipitated sulfur with CS_2, followed by benzene. A complexant citrate bicarbonate buffer is usually added now as a complexant to keep dissolved iron in solution. Coffin (1963) recommended a solution of 5% "hydrosulphite" in 0.15 *M* sodium citrate + 0.05 *M* citric acid at pH 4.25 and 50°C. Saunders (1959) demonstrated that neutral citrate alone was able to dissolve substantial quantities of Al from aluminium oxides and phosphate.

It has been reported that the dissolution of iron increases as the pH decreases (Coffin, 1963), but also that it does not (Mackenzie, 1954), and that dissolution is independent of the dithionite concentration between 2.5 and 20% (Coffin, 1963).

Solution of 1–4% dithionite have been used in 0.15 *M* sodium citrate + 0.5 *M* citric acid (Coffin, 1963), or in 0.175 *M* sodium citrate + 0.025 *M* citric acid (Tessier et al., 1979), or in 0.3 *M* sodium citrate + 1 *M* $NaHCO_3$ (Schwertmann, 1964), or in 1% trisodium citrate, or adjusted to pH 5.8 by NaOH (Mackenzie, 1954; Mitchell and Mackenzie, 1954), and in other less precisely recorded media.

It is reported that dithionite at room temperature dissolves "moderately reducible oxides" (Gupta and Chen, 1975) or amorphous iron oxides, and crystalline goethite, maghemite, magnetite, or lepidocrocite (McKeague and Day, 1966; Pawluck, 1972), to an extent that increases as these materials are more finely ground (McKeague and Day, 1966). However, at least part of the mobilized iron remains in a finely particulate suspension that also incorporates aluminium and silica (Follett et al., 1965).

Dithionite may dissolve aluminium too, from oxides and phosphates (Jarvis, 1986; Saunders, 1959), and can certainly release silica and aluminium from coprecipitates with iron oxide (Follett et al., 1965); however, it has little effect on ferriferous clays (Mehra and Jackson, 1960) or other clays (Mitchell and Mackenzie, 1954), although 2% dithionite is capable of reducing iron(III) in clay structures to iron(II) (Chen et al., 1987). It has been variously claimed that dithionite does not dissolve iron–organic complexes (Gibbs, 1973, 1977), or that it does dissolve all organic forms of Fe (Jeanroy et al., 1986b). Dithionite leaves a sample well bleached from iron oxides (Mitchell and Mackenzie, 1954), cleaner than does acid ammonium oxalate in the dark (Coffin, 1963). It may dissolve trace metal "precipitates" and sulfides (Tessier et al., 1979). A temperature of 50°C was quite adequate (Coffin, 1963) to dissolve "free" iron (i.e., all oxides), but the reagent did not dissolve pyrites, and only partially dissolved crystalline iron oxides or siderite (Coffin, 1963). At 70–80°C it has been used to dissolve "x-ray-amorphous" iron oxides and crystalline oxides (Miller and McFee, 1983; Smith and Mitchell, 1984) to release occluded trace metals. It also dissolves manganese oxides (which are more soluble than iron oxides) but not gibbsite (Tokashiki et al., 1986).

There is likely to be some readsorption of the released trace metals (Rendell et al., 1980), so various workers have washed the residues with 0.05 *M* $MgSO_4$ (Schwertmann, 1964), or 1 *M* NaCl after 0.05 *M* HCl, to dissolve precipitated sulfides (Mitchell and Mackenzie, 1954). However suitable for studies on iron oxides, dithionite is not ideal for work on trace metals because it is not available as a pure reagent and is not easily freed of small but significant quantities of trace metal impurities.

Q. Sodium Hypochloride (and Hypobromite)

Sodium hypochlorite is an oxidizing agent commonly used in histology as a clearing agent to dissolve nonstructural organic matter. In the present context it has been used to estimate "organically bound" trace metals (Gibbs, 1973, 1977; Shuman, 1979; Iyengar et al., 1981; Kuo et al., 1983; Miller et al., 1986a,b), presumably by destroying the organic groups or molecules to which they are attached. Troell (1931) found NaOBr very effective at dissolving amorphous organic matter, even in the presence of manganese dioxide. NaOCl has been used at pH 9.6 (Miller et al., 1986b), and in 6% solution (0.8 *M*) at pH 8.5 and 85°C (Kuo et al., 1983), where it mobilized the organically bound copper that was also dissolved by acid ammonium oxalate.

At 85°C and pH 8.5, NaOCl dissolved some 95% of soil organic matter, whereas H_2O_2 removed 65–80% (Gibbs, 1973, 1977). Miller et al. (1986a) assumed that NaOCl will mobilize less organic matter than pyrophosphate and that it does not dissolve manganese oxides. Because NaOCl liberates trace metals in their oxidized forms, these authors reduced them by dithionite, and then separated them from the remaining solids by dialysis. At high pH, NaOCl may lose dissolved metals (e.g., copper) by reprecipitation (Miller et al., 1986b).

R. Hydrogen Peroxide

Hydrogen peroxide is generally used, like NaOCl, to destroy organic matter and thereby release trace metals held by it. It is generally used at low pH, which may have other effects unrelated to the peroxide, but avoids the precipitation of metallic hydroxides. Note that commercial H_2O_2 usually contains added stabilizers, sometimes organic acids.

Various workers have extracted samples with 30% H_2O_2 at about pH 2 (in 0.01 *M* or 0.02 *M* nitric acid) once or twice, at room temperature (Rendell et al., 1980) or heated to 85°C or 100°C (Gupta and Chen, 1975; Shuman, 1979; Tessier et al., 1979; Wilber and Hunter, 1979; Harrison et al., 1981; Förstner et al., 1981; Hickey and Kittrick, 1984) to release organically bound trace metals. It has also been supposed to dissolve sulfides (Gupta and Chen, 1975; Tessier et al., 1979; Förstner et al., 1981; Miller et al., 1986b), and it has been shown to dissolve carbonates and oxides (Miller et al., 1986b), and reduce some oxides of manganese (Shuman, 1979; Miller et al., 1986a), possibly forming iron or aluminium oxalates (Follett et al., 1965; Shuman, 1979) or calcium oxalate (R.D. Macnicol and P.H.T. Beckett, in preparation) from the organic matter attacked. Robinson (1927) showed that some components of soil catalyze the attack of H_2O_2 on fibrous and resistant organic matter: Olmstead et al. (1930) and Alexander and Byers (1932) confirmed this and showed that small quantities of MnO_2 had a catalytic effect, although large quantities destroyed hydrogen peroxide: conversely, H_2O_2 partially dissolved manganese oxides (Miller et al., 1986a). Luoma and Jenne (1976) used 3% H_2O_2 and citrate buffer (0.2 *M* citric acid + 0.45 *M* sodium citrate) to remove trace metals adsorbed on organic matter and other materials.

Rendell et al. (1980) showed that a substantial part of the free copper in the reaction mixture may be readsorbed. To prevent this, various workers have added displacents to the peroxide, for example, 0.1 *M* sodium acetate (pH 4.0) (Gupta and Chen, 1975) or copper acetate (Iyengar et al., 1981), or have reextracted the reaction mixture at the end of the reaction with 1 *M* ammonium acetate in 6% HNO_3 (Gupta and Chen, 1975; Wilber and Hunter, 1979; Förstner et al., 1981) or 3.2 *M* ammonium acetate in 20% HNO_3 (Tessier et al., 1979; Harrison et al., 1981; Hickey and Kittrick, 1984), or 1 *M* $MgCl_2$ (Shuman, 1979).

S. Sodium Hydroxide

Sodium hydroxide is a general purpose extractant for soil organic matter, and 0.5 *M* sodium hydroxide has also been used to mobilize organically bound trace metals (Emmerich et al., 1982; Sposito et al., 1982). The dispersed organic matter may carry adsorbed/complexed trace metals into suspension with it (Miller et al., 1986b). NaOH removed more copper from pig manure than did 0.1 *M* pyrophosphate, and about the same amount as 0.1 *M* EDTA. Lund et al. (1985) used 0.5 *M* NaOH to mobilize organically bound trace metals, and showed that it also dissolved lead and zinc carbonates; and small proportions of cadmium, copper, and nickel carbonates, although not the sulfides of these elements.

Hot 1.25 *M* NaOH (75 °C) dissolves kaolinite without affecting the Al substitution of other minerals present (Kämpf and Schwertmann, 1982); hot 0.5 *M* NaOH has been reported to dissolve gibbsite and poorly ordered aluminosilicates, but other silicates only slowly, and to dissolve amorphous alumina and silica but not to attack crystalline clays (Hashimoto and Jackson, 1960).

Strong NaOH solutions extract aluminium or silica from minerals. Thus, boiling 5 *M* NaOH has been reported to dissolve kaolinite and gibbsite (Kämpf and Schwertmann, 1982). These authors reported that NaOH tended to dissolve silica or aluminium preferentially from iron oxides so that it increased the crystallinity of Al goethite and reduced its Al substitution (e.g., from 30 to 10 mol%), although it had no effect on Al haematite; it may be used to clean minerals without affecting their goethite/haematite ratio, provided there is at least 0.2 *M* silica in the solution to prevent recrystallization. Norrish and Taylor (1961) reported that it dissolved clays to leave iron oxides. According to Tokashiki et al. (1986), 5 *M* NaOH at 90°C tends to increase the crystallinity of weakly crystalline Mn Compounds. It may crystallize amorphous manganese oxides into birnessite, but it does not dissolve birnessite or lithiophorite, nor does it increase their crystallinity.

T. Sodium Carbonate

Sodium carbonate might be used as a weaker and better buffered alkali than sodium hydroxide, for mobilizing and/or hydrolyzing organic matter, but appears not to have been used in this way.

Hot 5% solutions (0.47 *M*) of sodium carbonate have been used to dissolve gibbsite, allophane, and poorly ordered aluminosilicates (probably from the surfaces of clay plates) (Follett et al., 1965; Smith and Mitchell, 1984), and cold 5% Na_2CO_3 to remove the last. Iron compounds or crystalline clays are not dissolved, although the removal of noncrystalline aluminosilicates may make the iron oxides more easily accessible to other reagents (Follett et al., 1965).

U. Nitric Acid

As well as being an acid, nitric acid is an oxidizing agent. It has been used to analyze "total" trace metals in sewage sludges ("strong acids," below), although it can be shown that it does not always dissolve as much of them as do HF or $HClO_4$. If used in a sequence of extractants, HNO_3 usually comes last, to dissolve the groups of compounds, often supposed to be sulfides (Emmerich et al., 1982), that are resistant to the preceding extractants.

Thus, 1 *M* HNO_3 has been reported as dissolving or displacing:

Sulfides (not dissolved by EDTA, KF, pyrophosphate) (Stover et al., 1976);
Sulfides and a range of "acid-soluble" components (Miller and McFee, 1983);
"Strongly adsorbed" cations not previously displaced by copper acetate (Soon and Bates, 1982);
"Adsorbed metals" (Schalscha et al., 1982);
Adsorbed cations, unspecified "precipitates," and some organically bound cations (Silviera and Sommers, 1977).

Lund et al. (1985) reported that 2 *M* HNO_3 dissolved nickel, lead, and zinc sulfides; 4 *M* HNO_3 at 80°C dissolved cadmium, nickel, lead, and zinc sulfides, and partly dissolved copper sulfides. Sposito et al. (1982) assumed that 4 *M* HNO_3 at 80°C dissolved the sulfides that are not dissolved by EDTA or NAOH.

V. Hydrochloric Acid

Solutions of 0.1 *M* hydrochloric acid have been used to extract "exchangeable" (Shuman, 1979) or "H-exchangeable" (Luoma and Jenne, 1976) cations, and have been shown to displace cadmium, cobalt, and zinc freshly adsorbed onto iron and manganese oxides, carbonates, or decomposing organic matter (Luoma and Jenne, 1976); 0.5 *M* HCl has been used to extract "acid-soluble" copper, and has been shown to dissolve copper oxide and basic copper carbonate, but only a little copper sulfide (Luoma and Jenne, 1976). Mitchell and Mackenzie (1954) reported that 0.05 *M* HCl dissolves freshly precipitated iron sulfide.

Chao and Zhou (1983) used HCl at various concentrations, and found that it dissolved only a little of crystalline iron oxides at room temperature: dissolution was in the order amorphous > magnetite > goethite > hematite.

W. Strong Acids

The "residual" trace metals remaining after the extractants listed previously, or the "total" quantity of trace metals, have been estimated by digestion in strong acids.

Many combinations of acids have been used, for example, molar 1 *M* HNO_3 (Silviera and Sommers, 1977); 4 *M* HNO_3 (Emmerich et al., 1982; Sposito et al., 1982); concentrated HNO_3 at 180°C (Förstner et al., 1981); concentrated HNO_3 and H_2O_2 (Miller and McFee, 1983); HNO_3 and $HClO_4$ (Schalscha et al., 1982); HNO_3 and HF (Wilber and Hunter, 1979; Harrison et al., 1981); HNO_3, HF, and $HClO_4$ (Gupta and Chen, 1975; Cottenie et al., 1979); HF and $HClO_4$ (Tessier et al., 1979; Hickey and Kittrick, 1984); HF (Sims and Patrick, 1978; McLaren et al., 1986; Miller et al., 1986a); $HClO_4$ (Cottenie et al., 1979); and aqua regia (Berrow and Stein, 1983; Miller et al., 1986a), etc.

Agemian and Chau (1976) showed that aqua regia was slightly more effective at mobilizing metals from sediments than boiling HNO_3–$HClO_4$ (1:1). Oliver

(1973) extracted riverine sediments with 4 *M* HNO_3 + 0.07 *M* HCl at 70–90°C and showed that the quantities of trace metals extracted showed a roughly constant shortfall from the quantities dissolved by HF + HNO_3 + $HClO_4$. He inferred that the former probably dissolved most forms of the trace metals, but not crystalline minerals. Similarly Berrow and Stein (1983) showed that strong acids or aqua regia did not dissolve all of the trace metals present, as analyzed by dc arc spectrophotometer. They also showed that aqua regia was not necessarily more effective than single acids, and commented that the efficiency of strong acids might depend very much on how finely the samples were ground.

IV. Discussion

The range of publications on extractants, when juxtaposed, is not entirely encouraging. There are few extractants to which contradictory and inconsistent effects have not been attributed by different workers.

Some discrepancies are the result merely of differences in terms, but others are fundamental. Obviously the problem can be circumvented by defining categories of trace metal combination in purely operational terms, for example, "HCl soluble," "pyrophosphate soluble," with precise specifications of the extractants and procedures involved. Some workers do this, and it may be sufficient for simple comparisons of samples, or of different stages in sludge transformations. However, this solution to the problem rather ignores the original purpose of the extractants, which was to provide a means for elucidating the chemistry of trace metals in sludges or soils in order to explain, and ultimately to predict, their transformations.

It is also all too clear that more checks are needed on what different extractants do in fact extract. Much of the very limited work on the solution of defined compounds is affected by a separate problem—that the ease of dissolution of a given compound is very sensitive to its physical form. The effect of varying crystallinity on solubility has been extensively discussed in the case of oxides, but not for other groups of compounds. Solubility is also affected both by whether the material is coarsely or finely divided and by whether it is intimately mixed, or surrounded by, less soluble materials. It would seem that checks must be supported by XRD measurements, not only to establish the crystallinity of the compound being extracted, but also to follow its disappearance by dissolution.

Whether or not it is, a priori, possible to devise extractants capable of distinguishing all the various categories of trace metal combinations outlined on pp. 146–149, it would seem clear that too many of the extractants in common use have a double or mixed action. For example acetic acid offers a displacing cation, a complexing anion, and a hydrolytic cation simultaneously.

The main effects of specific extractants are

1. Cation exchange or displacement by:
   a. simple cations like Ca and Mg,

   b. less simple cations like Cu or Pb,
   c. protons, or
2. Complexation by complexants competing against sites in the soil or sludge:
   a. acetate,
   b. citrate, oxalate, or fluoride,
   c. EDTA or pyrophosphate, or
3. Dispersion/peptization by
   a. pyrophosphate,
   b. hydroxide, or
4. Hydrolysis by strong acids or alkalis; or
5. Redox reactions by reagents such as
   a. hydroxylamine or dithionite, or
   b. NaOCl, H_2O_2, etc.

It would be much more informative if these processes could be distinguished, for example, by using the two extractants sodium acetate and copper chloride separately in place of copper acetate alone.

Pyrophosphate is a particularly complicated example. It has both complexant and dispersant effects, of which the latter depends in part on the former. It was originally used as an extractant for organic matter, and in this case the double effect did not matter. However, when it is used as an extractant for organically bound trace metals, it is not clear whether it is the uncentrifuged suspension or the centrifuged supernatant from the extract that is to be analyzed (pp. 157–159). The suspension is likely to include trace metals held on suspended organic matter, of which as much as half may be lost by centrifuging. If the suspension is to be centrifuged before analysis, then presumably the pellet must be restored to the pyrophosphate residue, to be reextracted by the next extractant. It is not clear that this is always done. This is just one example; there are similar problems with other extractants.

The reimmobilization of displaced iron or trace metals is a further problem. It is unlikely to be serious in the stages in which the extractant (e.g., of displacing cations or mild complexants) operates by equilibrating with a category of trace metals in the sample but otherwise does not alter the sample. It is potentially more serious when a strong complexant actually modifies the solid phase by removing one component, or destroys one component to release the trace metals it holds or occludes. Such processes may introduce a range of new adsorption sites or amorphous precipitates. Rendell et al. (1980) reported that sediment cores under extraction by dilute HCl, hydroxylamine, ammonium acetate, or dithionite reagents absorbed significant quantities of the copper that had been added to the extractants at the beginning, but not from EDTA extractants. Nor can it be assumed that trace metals released on the destruction of organic matter by oxidizing agents will not be immobilized on other components.

Complexants may be added to extractants to retain the displaced trace metals in solution. Thus Jarvis (1984) added 1 *M* ammonium acetate to ammonium oxalate and dithionite extractants, Wilber and Hunter (1979) added it to 6% nitric acid, and Tessier et al. (1979), Harrison et al. (1981), and Hickey and Kittrick (1984) added 3.2 *M* ammonium acetate in 20% nitric acid to their H_2O_2 extractants. Gupta and Chen (1975) found that 1 *M* ammonium acetate was more effective at preventing readsorption than 0.01 *M* nitric acid. McLaren et al. (1986) washed the residue from each extractant with 0.05 *M* calcium chloride, and Miller et al. (1986a) with 0.1 *M* calcium nitrate. Luoma and Jenne (1976) washed their dithionite residue with 6 *M* sodium nitrate, and Schwertmann (1964) washed ammonium oxalate residues with 0.05 *M* magnesium sulfate.

However, it cannot be assumed that the added complexant does not have extractant effects too, as is the case for citrate added to the dithionite extractant (p. 154). There is no simple answer. At least an added complexant should be substantially less aggressive than the extractant to which it is added. It might be better to explore the use of a leaching apparatus to remove displaced ions relatively rapidly as they are displaced, rather than leaving them in contact with the sample during 6–16 h of shaking.

There are also minor points. Drying a sample (particularly by heating), or storing a moist sample, are both likely to alter the forms of combination of trace metals, not only those held on organic ligands likely to be affected by microbial activity, but also those likely to pass from amorphous to crystalline forms on heating or storage.

Thus, Miller et al. (1986b) compared freeze-dried and oven-dried pig manure: the quantities of copper dissolved from the two samples by different extractants were not even in the same order. Miller et al. (1986b) and Jarvis (1984) also reported effects of drying. Silviera and Sommers (1977) found that as well as affecting the extractability of trace metals to different extractants, drying also affected the interactions of sludge and soil, at least in the early stages. Martin and Reeve (1957) found that even solvent-drying affected the solution of organic matter. It should not be assumed, without trial, that a dried and ground sample suffers no further changes on storage; quite small changes in the crystallinity of iron and other oxides may have a considerable effect on their extractability and their ability to release trace metals. Also, some time is required for a dried sample to become fully rewetted, and it cannot be assumed that its trace metals will return to their original forms of combination (Le Riche and Weir, 1963; Mackenzie, 1954).

Samples of sludge, or of soil recently treated with sludge or slurry, are often in a state between fully anaerobic and fully oxidized conditions. Very few workers have considered that the preparation of samples, and their manipulation during a sequence of extractions, may actually accelerate changes in compounds originally formed at low redox potentials (e.g., sulfides or iron(II) compounds) and metastable under aerobic conditions, unless pains are taken to prevent it. It is likely that the rate and extent of oxidation may vary between samples in quite unpredictable and inconsistent ways. Gupta and Chen (1975) and Sims and Patrick (1978) performed extractions under nitrogen or argon.

As has been suggested, it is surprising that so few workers (and the author of this paper only fortuitously) have removed oils and waxes from samples of sludge or sludge-treated soils before applying a sequence of aqueous extractants. It is unlikely that a preliminary cleaning with a nonpolar solvent will remove any substantial quantity of the trace metals normally dissolved by aqueous extractants, but it could increase their accessibility to other extractants.

There has been surprisingly little examination of the use of ultrasonic dispersion to open up clumps of organic matter, and to disperse mixtures of colloidal material and secondary precipitates.

In some cases at least it is clear that the composition of an extractant, or its pH, may be significantly modified by the samples to be analyzed. For example, Miller et al. (1986a) reported that the hydroxylamine hydrochloride reagent was less effective after pyrophosphate (pH 10) than before; it appears that the material being analyzed was sufficiently strongly buffered to raise the pH of the hydroxylamine from pH 2.2. to 3.3. Luoma and Jenne (1976) reported that the carbonates in their samples partially neutralized an acid reagent. In both these examples, the extent of these effects could have varied between samples in ways unrelated to what was being measured.

In conclusion, it must be emphasized:

1. There is a real and urgent need for selective extractants to distinguish and estimate the different forms of combination of trace metals that occur when sludge and soil interact.
2. The development of such extractants requires more effort than it has so far received.

Acknowledgments

Acknowledgments are due to Ann Sykes, whose work stimulated the preparation of this review, and to Roger MacNicol for pointing out the pyrophosphate paradox.

References

Adams, F., C. Burmester, N.V. Hue, and F.L. Long. 1980. A comparison of column displacement and centrifuge methods for obtaining soil solutions. *Soil. Sci. Soc. Am. J.* 44:733–735.

Agemian, H., and A.S.Y. Chau. 1976. Evaluation of extraction techniques for the determination of metals in aquatic sediments. *Analyst* 101:761–767.

Alexander, L.T., and H.G. Byers, 1932. A critical laboratory review of methods of determining organic matter and carbonates in soil. U.S. Dept. of Agriculture, *Tech. Bull. No. 317*.

Ball, D.F., and P. Beaumont. 1972. Vertical distribution of extractable iron and aluminium in soil profiles from a brown earth-peaty podzol association. *J. Soil. Sci.* 23:298–308.

Baldwin, A., T.A. Brown, P.H.T. Beckett, and G.E.P. Elliott. 1983. The forms of combination of Cu and Zn in digested sewage sludge. *Water Res.* 17:1935–1944.

Bascomb, C.L. 1968. Distribution of pyrophosphate-extractable iron and organic carbon in soils of various groups. *J. Soil Sci.* 19:251–268.

Bergman, S.C., C.J. Ritter, E.E. Zamierowski, and C.R. Cothern. 1979. The use of zonal centrifugation in delineating trace element distributions in sewage sludges from the Dayton, Ohio area. *J. Environ. Qual.* 8:416–422.

Berrow, M.L., and W.M. Stein. 1983. Extraction of metals from soils and sewage sludges by refluxing with aqua regia. *Analyst* 108:277–285.

Borggard, O.K. 1976. Selective extraction of amorphous iron oxides by EDTA from a mixture of amorphous iron oxide, goethite and haematite. *J. Soil Sci.* 27:478–486.

Bracewell, J.M., A.S. Campbell, and B.D. Mitchell. 1970. An assessment of some thermal and chemical techniques used in the study of the poorly-ordered alumino-silicates in soil clays. *Clay Miner.* 8:325–335.

Bradford, G.R., A.L. Page, L.J. Lund, and W. Olmstead. 1975. Trace element concentrations of sewage treatment plant effluents and sludges; their interactions with soils and uptake by plants. *J. Environ. Qual.* 4:123–127.

Bremner, J.M., and H. Lees. 1949. Studies on soil organic matter. II. The extraction of organic matter from soil by neutral reagents. *J. Agric. Sci. (Cambridge)* 39:274–279.

Brown, A.L., J. Quick, and J.L. Eddings. 1971. A comparison of analytical methods for soil zinc. *Soil Sci. Soc. Am. Proc.* 35:105–107.

Campbell, D.J., and P.H.T. Beckett. 1988. The soil solution in a soil treated with digested sewage sludge. *J. Soil Sci.* 39:283–298.

Chao, T.T. 1972. Selective dissolution of manganese oxides from soils and sediments with acidified hydroxylamine hydrochloride. *Soil Sci. Soc. Am. Proc.* 36:764–768.

Chao, T.T., and L. Zhou. 1983. Extraction techniques for selective dissolution of amorphous iron oxides from soils and sediments. *Soil Sci. Soc. Am. J.* 47:225–232.

Chen, S.Z., P.F. Low and C.B. Roth. 1987. Relation between potassium fixation and the oxidation state of octahedral iron. *Soil Sci. Soc. Am. J.* 51:82–90.

Coffin, D.E. 1963. A method for the determination of free iron in soils and clays. *Can. J. Soil Sci.* 43:7–17.

Cottenie, A., R. Camerlynck, M. Verloo, and A. Dhaese. 1979. Fractionation and determination of trace elements in plants, soils and sediments. *Pure Appl. Chem.* 52:45–53.

Dion, H.C. 1944. Iron oxide removal from clays and its influence on base-exchange properties and x-ray diffraction patterns of the clays. *Soil Sci.* 58:411–424.

de Endredy, A.S. 1963. Estimation of free iron oxides in soils and clays by a photolytic method. *Clay Miner. Bull.* 5:209–217.

Eaton, F.M., R.B. Harding, and T.J. Ganje. 1960. Soil solution extracts at tenth-bar moisture percentage. *Soil Sci.* 90:253–258.

Emmerich, W.E., L.J. Lund, A.L Page, and A.C. Chang. 1982. Solid phase forms of heavy metals in sewage sludge-treated soils. *J. Environ. Qual.* 11:178–181.

Fletcher, P., and P.H.T. Beckett. 1987. The chemistry of heavy metals in digested sewage sludge. 2. Heavy metal complexation with soluble organic matter. *Water Res.* 21: 1163–1172.

Follett, E.A.C., W.J. McHardy, B.D. Mitchell, and B.F.L. Smith. 1965. Chemical dissolution techniques in the study of clays. Pt. I and II. *Clay Miner.* 6:23–34; 35–43.

Förstner, U.W., K. Calmano, H. Conrad, H. Jaksch, C. Schimkus, and J. Schoer. 1981. Chemical speciation of heavy metals in solid waste materials (sewage sludge, mining wastes, dredged materials, polluted sediments) by sequential extraction. *Proc. Int. Conf. Heavy Metals Environ*, WHO/EED, pp. 698–704.

Gibbs, R.J. 1973. Mechanisms of trace metal transport in rivers. *Science* 180:71–73.

Gibbs, R.J. 1977. Transport phases of transition metals in the Amazon and Yukon rivers. *Geol. Soc. Am. Bull.* 88:829–843.

Gupta, S.K., and Chen. 1975. Partitioning of trace elements in selective chemical fractions of nearshore sediments. *Environ. Lett.* 10:129–158.

Hamblin, A.P., and A.M. Posner. 1979. The use of acetylacetone as a selective extractant of organically bonded metals in soils. *J. Soil Sci.* 30:175–181.

Harrison, R.M., D.P.H. Laxe, and S.J. Wilson. 1981. Chemical associations of lead, cadmium, copper and zinc in street dusts and roadside soils. *Environ. Sci. Technol.* 15:1378–1383.

Hashimoto, I., and M.L. Jackson. 1960. Rapid dissolution of allophane and kaolinite-halloysite after dehydration. *Clays and Clay Miner.* 7:102–13.

Hickey, M.G., and J.A. Kittrick. 1984. Chemical partitioning of Cd, Cu, Ni and Zn in soils and sediments containing high levels of heavy metals. *J. Environ. Qual.* 13:372–377.

Himes, F., and S. Barber. 1957. Chelating ability of soil organic matter. *Soil Sci. Soc. Am. Proc.* 21:368–373.

Hodgson, J.F. 1963. Chemistry of the micronutrient elements in soils. *Adv. Agron.* 15:119–159.

Iyengar, S.S., D.C. Martens, and W.P. Miller. 1981. Distribution and plant availability of soil zinc fractions. *Soil Sci. Soc. Am. J.* 45:735–739.

Jarvis, S.C. 1984. The forms of occurrences of Mn in some acidic soils. *J. Soil Sci.* 35:421–429.

Jarvis, S.C. 1986. Forms of Al in some acid permanent grassland soils. *J. Soil Sci.* 37:211–222.

Jeanroy, E., and B. Guillet. 1981. The occurrence of suspended ferruginous particles in pyrophosphate extracts of some soil horizons. *Geoderma* 26:95–106.

Jeanroy, E., B. Guillet, and R. Ortiz. 1986b. Evaluation of iron forms by chemical extractants: applications to brunified and podzolic soils. *Science du Sol*, pp. 137–136.

Jeanroy, E., B. Guillet, P. Delcroix, and C. Janot. 1986a. Soil iron forms: a comparison between chemical methods and Mössbauer spectroscopy. *Science du Sol*, pp. 135–136.

Kämpf, N., and U. Schwertmann. 1982. The 5 *M* NaOH concentration treatment for iron oxides in soils. *Clays Clay Miner.* 30:401–408.

Kuo, S.P., P.E. Heilman, and A.S. Baker. 1983. Distribution and forms of Cu, Zn, Cd, Fe and Mn in soils near a copper smelter. *Soil Sci.* 135:101–109.

Le Riche, H.H., and A.H. Weir. 1963. A method of studying trace elements in soil fractions. *J. Soil Sci.* 14:225–235.

Lester, J.N., R.M. Sterritt, and P.W.W. Kirk. 1983. Significance and behaviour of heavy metals in waste water treatment processes. II. Sludge treatment and disposal. *Sci. Total Environ.* 30:45–83.

Lindsay, W.L. 1979. *Chemical Equilibria in Soils.* Wiley, New York.

Lindsay, W.L., and W.A. Norvell. 1969. Equilibrium relationships of Zn^{++}, Fe^{+++}, Ca^{++} and H^{+} with EDTA and DTPA in soils. *Soil Sci. Soc. Am. Proc.* 33:62–68.

Loveland, P.J., and P. Bullock. 1976. Chemical and mineralogical properties of known podzolic soils in comparison with soils of other groups. *J. Soil Sci.* 32:523–540.

Loveland, P.J., and P. Digby. 1984. The extraction of Fe and Al by 0.1 *M* pyrophosphate solutions: a comparison of some techniques. *J. Soil Sci.* 35:243–250.

Lund, L.J., G. Sposito, and A.L. Page. 1985. Project Summary–Determination and prediction of chemical forms of trace metals in sewage sludge and sludge-amended soils. EPA/600/52-85/053, U.S. Environmental Protection Agency, Water Engineering Research Laboratory.

Luoma, S.N., and E.A. Jenne. 1976. Estimating bio-availability of sediment-bound trace metals with chemical extractants. *Trace Subst. Environ. Health* 10:343–351.

Mackenzie, R.C. 1954. Free iron-oxide removal from soils. *J. Soil Sci.* 5:167–172.

Mandal, L.N., and B. Mandal. 1987. Transformation of zinc fractions in rice-soils. *Soil Sci.* 143:205–212.

Martin, A.E., and R. Reeve. 1957. Chemical studies of podzolic illuvial horizons. II. The use of acetyl acetone as extractant of translocated organic matter. *J. Soil Sci.* 8:279–286.

Mattigod, S.V., A.L. Page, and I. Thornton. 1986. Identification of some trace metal minerals in a mine-waste contaminated soil. *Soil Sci. Soc. Am. J.* 50:254–258.

McBride, M.B. 1981. Forms and distribution of Cu in solid and solution phases of soil. In J.F. Loneragan (ed.), *Copper in Soils and Plants*, pp. 24–45. Academic Press, Sydney, Australia.

McKeague, J.A. 1967. An evaluation of 0.1 *M* pyrophosphate and pyrophosphate-dithionite in comparison with oxalate as extractants of accumulation products in podzols and some other soils. *Can. J. Soil Sci.* 47:95–99.

McKeague, J.A., and J.H. Day. 1966. Dithionite- and oxalate-extractable Fe and Al as aids in differentiating various classes of soils. *Can. J. Soil Sci.* 46:13–22.

McLaren, R.G., and D.V. Crawford. 1973. Studies on soil copper. I. The fractionation of Cu in soils. *J. Soil Sci.* 24:172–181.

McLaren, R.G., D.M. Lawson, and R.S. Swift. 1986. The forms of cobalt in some Scottish soils as determined by extraction and isotopic exchange. *J. Soil Sci.* 37:223–234.

Mehra, O.P., and M.L. Jackson. 1960. Iron oxide removal from soils and clays by a dithionite-citrate system buffered with sodium bicarbonate. *Proc. 7th Natl. Conf. Clays Clay Miner.* 9:317–327.

Miller, W.P., and W.W. McFee. 1983. Distribution of Cd, Zn, Cu and Pb in soils of industrial north western Indiana. *J. Environ. Qual.* 12:29–33.

Miller, W.P., D.C. Martens, and L.W. Zelazny. 1986a. Effect of sequence in extraction of trace metals from soils. *Soil Sci. Soc. Am. J.* 50:558–560.

Miller, W.P., D.C. Martens, L.W. Zelazny, and E.T. Kornegay. 1986b. Forms of solid-phase copper in copper-enriched swine manure. *J. Environ. Qual.* 15:69–72.

Ministry of Agriculture, Fisheries and Food. 1981. *The Analysis of Agricultural Materials*. Ref. Book 427, Ministry of Agriculture, Fisheries and Food, London.

Mitchell, B.D., and R.C. MacKenzie. 1954. Removal of free iron oxide from clays. *Soil Sci.* 77:173–184.

Murthy, A.S.P. 1982. Zinc fractions in wetland rice soils and their availability to rice. *Soil Sci.* 133:150–154.

Neilsen, D., P.B. Hoyt, and A.F. MacKenzie. 1986. Distribution of soil Zn fractions in British Columbia interior orchard soils. *Can. J. Soil Sci.* 66:445–454.

Norrish, K., and R.M. Taylor. 1961. The isomorphous replacement of iron by aluminium in soil goethites. *J. Soil Sci.* 12:294–306.

Oliver, B.G. 1973. Heavy metal levels of Ottawa and Rideau river sediments. *Environ. Sci. Technol.* 7:135–137.

Olmstead, L.B., L.T. Alexander, and H.E. Middleton. 1930. A pipette method of mechanical analysis of soils based on improved dispersion procedures. U.S. Dept. of Agriculture, *Tech. Bull. 170*.

Pawluck, S. 1972. Measurement of crystalline and amorphous iron removal in soils. *Can. J. Soil Sci.* 52:119–123.

Rappaport, B.D., D.C. Martens, T.W. Simpson, and R.B. Reneau. 1986. Prediction of available zinc in sewage sludge-amended soils. *J. Environ. Qual.* 15:133–136.

Rendell, P.A., G.E. Batley, and A.J. Cameron. 1980. Adsorption as a control of metal concentrations in sediment extracts. *Environ. Sci. Technol.* 14:314–318.

Rich, C.I., and S.S. Obenshain. 1955. Chemical and clay mineral properties of a red-yellow podzolic soil derived from mica-schist. *Soil Sci. Soc. Am. Proc.* 19:334–339.

Robinson, W.O. 1927. The determination of organic matter in soil by means of hydrogen peroxide. *J. Agric. Res. (Cambridge).* 34:339–356.

Saunders, W.M.H. 1959. Aluminium extracted by neutral citrate-dithionite reagent. *Nature (London)* 184:2037.

Schalscha, E.B., M. Morales, I. Ahumada, T. Schirado, and P.F. Pratt. 1980. Fractionation of Zn, Cu, Cr and Ni in waste water solids and in soil. *Agrochimica* 24:361–368.

Schalscha, E.G., M. Morales, I. Vergara, and A.C. Chang. 1982. Chemical fractionation of heavy metals in waste-water affected soils. *J. Water Pollut. Control Fed.* 54:175–180.

Schnitzer, M., and J.R. Wright. 1956. Notes on the extraction of organic matter from the B horizon of a podzol soil. *Can. J. Soil Sci.* 36:511–512.

Schwertmann, U. 1964. The differentiation of iron oxides in soils by a photochemical extraction with acid ammonium oxalate. *Z. Pflanzenernahr. Dueng. Bodenk.* 105:194–201.

Scott, A.D., and L.F. Welsh. 1961. Release of non-exchangeable soil potassium during short periods of cropping and sodium tetraphenyl boron extraction. *Soil Sci. Soc. Am. Proc.* 25:128–130.

Scott, N.M., and G. Anderson. 1976. Sulphur, carbon and nitrogen contents of organic fractions from acetyl acetone extracts of soil. *J. Soil Sci.* 27:324–330.

Sedberry, J.E., and C.N. Reddy. 1976. The distribution of zinc in selected soils in Indiana. *Commun. Soil Sci. Plant Anal.* 7:787–795.

Shuman, L.M. 1979. Zn, Mn and Cu in soil fractions. *Soil Sci.* 127:10–17.

Silviera, D.J., and L.E. Sommers. 1977. Extractability of Cu, Zn, Cd and Pb in soils incubated with sewage sludge. *J. Environ. Qual.* 6:47–52.

Sims, J.L., and W.H. Patrick. 1978. The distribution of micronutrient cations in soil under conditions of varying redox potential and pH. *Soil Sci. Soc. Am. J.* 42:258–262.

Smith, B.F.L., and B.D. Mitchell. 1984. Characterisation of x-ray amorphous material in a Scottish soil by selective chemical techniques. *Clay Miner.* 19:737–744.

Soon, Y.K., and T.E. Bates. 1982. Chemical pools of Cd, Ni and Zn in polluted soils and some preliminary indications of their availability to plants. *J. Soil Sci.* 33:477–488.

Sposito, G., L.J. Lund, and A.C. Chang. 1982. Trace metal chemistry in arid-zone field soils amended with sewage sludge I: Fractionation of Ni, Cu, Zn, Cd and Pb in solid phases. *Soil Sci. Soc. Am. J.* 46:260–264.

Stover, R.C., L.E. Sommers, and D.J. Silviera. 1976. Evaluation of metals in waste-water sludge. *J. Water Poll. Control Fed.* 48:2165–2175.

Strachan, S.D., D.W. Nelson, and L.E. Sommers. 1983. Sewage sludge components soluble with non-aqueous solvents. *J. Environ. Qual.* 12:69–74.

Tamm, O. 1922. [Method for the estimation of the inorganic components of the gel complex in soils.] *Medd. Skogsförsokanst.* 19:387–404.

Tessier, A., P.G.C. Campbell, and M. Bisson. 1979. Sequential extraction procedure for the speciation of particulate trace metals. *Anal. Chem.* 51:844–851.

Tiller, K.G., J.L. Honeysett, and M.P.C. de Bries. 1972. Soil zinc and its uptake by plants. II. Soil chemistry in relation to prediction of availability. *Aust. J. Soil Res.* 10:165–182.

Tokashiki, Y., J.B. Dixon, and D.C. Golden. 1986. Manganese oxide analysis in soils by combined x-ray diffraction and selective dissolution methods. *Soil Sci. Soc. Am. J.* 50:1079–1084.

Towe, K.M., and W.F. Bradley. 1967. Mineralogical constitution of colloidal "hydrous ferric oxides." *J. Coll. Interface Sci.* 24:384–392.

Troell, E. 1931. The use of sodium hypobromite for the oxidation of organic matter in the mechanical analysis of soils. *J. Agric. Sci. (Cambridge)* 21:476–484.

Viets, F.G. 1962. Chemistry and availability of micro-nutrients in soils. *J. Agric. Food Chem.* 10:174–178.

Viets, F.G., and W.L. Lindsay. 1973. Testing soils for Zn, Cu, Mn and Fe. In L.M. Walsh and J.D. Beaton (eds.), *Soil Testing and Plant Analysis*. Soil Science Society of America, Inc., Madison, Wisconsin.

Wang, C., and P.A. Schuppli. 1986. Determining ammonium oxalate–extractable Si in soils. *Can. J. Soil Sci.* 66:751–755.

Wilber, W.G., and J.V. Hunter. 1979. Distribution of metals in street sweepings, stormwater solids and urban aquatic sediments. *J. Water Pollut. Control Fed.* 51:2810–2822.

Using Soil Survey Data for Quantitative Land Evaluation

J. Bouma

I. Introduction 177
II. Soil Survey Interpretation and Land Evaluation 179
A. Qualitative and Semiquantitative Procedures 179
B. Quantitative Procedures 181
III. Modern Land Use Questions 185
A. Agriculture Versus Environment 185
B. Problem Analysis 186
IV. Innovative Uses of Soil Survey Data 192
A. Sampling and In Situ Measurements 192
B. Obtaining Basic Data for Simulation Modeling 195
C. Modeling Water Regimes in Soils with Macropores 201
D. Structural Phases of Soil Series 205
E. Spatial Interpolation 207
V. Future Developments 209
References 211

I. Introduction

Soil survey activities in many countries have reached a crucial phase. Standard country wide surveys either have been completed or will be completed within the near future. This development implies that more attention can be paid to soil survey interpretations and applications. The opportunity to emphasize use of soil survey data is particularly timely because of a change in the type of questions being asked by users of soil survey data and because of the rapid advances in information technology. Questions have become more specific, and quantitative answers are increasingly required.

Soil surveys can only start after the development of theories of soil genesis and soil classification that explain the occurrence of particular soil types in different landscapes. Soil classification systems are still being perfected, but the major thrust of research has moved to other areas such as interpretation, after the publi-

Advances in Soil Science, Volume 9

cation of the comprehensive Soil Taxonomy (1975) and the FAO-UNESCO classification system (1987). Modern soil surveys contain interpretations for a wide variety of uses, defining relative limitations or suitabilities in a number of empirical classes, essentially based on expert judgment.

Introduction of the FAO framework for land evaluation in 1976 has been an important step toward improving interpretations by not only defining suitability classes, as such, but also determining specific, well defined land qualities on which these suitabilities are to be based.

More recently, quantification has been further increased by introducing computer simulation techniques to define land qualities for a variety of climatic and management-induced conditions. This offers the attractive possibility of producing a wide variety of interpretations that in principle can address any particular question. This development is particularly fortunate because modern users of soil survey information are increasingly professional. Their questions are becoming more specific, and they are often more interested in a quantitative listing of options available, rather than in a broad final judgment in terms of a suitability for a given purpose. They prefer to make their own selections.

Computer simulation of soil processes is only realistically possible when basic soil data are readily available. Use of soil information systems is necessary in this context. Obtaining basic data that are relevant for field conditions is one of the major challenges that soil science will face in the years ahead, and this aspect receives considerable emphasis here. Use of modern information technology to cater toward specific questions of individual users represents an important modern direction in soil survey work.

Aside from the effect of users' demands and of technological development within soil science itself, soil physical and chemical research increasingly emphasizes spatial variability aspects because modern environmental and agricultural questions focus on *areas* of land. Legal aspects require quantitative and reproducible answers, preferably expressed in statistical terms. A conventional statement such as "the iron content is 4.0%" is likely to become less acceptable than "the content of iron has a probability of 80% of being higher than 4.0%." Aside from considering spatial variability, we must also critically examine our theoretical concepts. Field studies reveal that soil structures and associated flown patterns of solutes are far more complicated than those in homogeneous porous media. Even though awareness of variability aspects and of complex flow systems in the field is growing, increasing the usefulness of soil survey data for realistic simulation models that will be used for defining various land qualities still represents a major challenge.

Pedology began with observations of soil features in the field that had been there for thousands of years but that had not yet been "seen" by anybody. The clear need to channel our by now vast knowledge of soils into classification systems and data banks is acknowledged. However, we are in trouble when not enough time is left to continually "see" new features and to develop new ideas and concepts. As soon as such developments are considered as a nuisance because they require adaption of THE SYSTEM, we are really in trouble. This chapter is intended to stimulate discussions about future use of soil survey data and future soil survey activities covering characterization of areas of land.

II. Soil Survey Interpretation and Land Evaluation

A. Qualitative and Semiquantitative Procedures

Soil maps have been widely used for interpretive purposes by defining relative suitabilities or limitations of various soil types for different land uses. Such interpretations are essentially based on practical experience of soil surveyors as gained from their extensive contact with farmers, agricultural agents, and colleagues. As a consequence, these interpretations are quite valuable because they usually reflect the current level of knowledge.

Once suitabilities or limitations for a given type of land use have been defined, they can be shown on soil maps because of their direct association with delineated areas on the soil map according to the map legend. As the need for more specific interpretations increased, two modifications have been made. *First*, emphasis was placed on the definition of different so-called diagnostic factors that determine suitabilities or limitations. These can be land characteristics (easy to measure in the field) or land qualities (to be calculated) (e.g., FAO, 1976, 1983; Bouma and van Lanen, 1987). Together, these diagnostic features determine soil suitability by matching crop or ecological *requirements* with the *properties* of a particular land unit. *Second*, a more specific definition of land utilization types (LUT) was required for expressing the major importance of socioeconomic aspects when using land. Each land evaluation is to be made for a specific LUT.

A detailed discussion of land evaluation procedures is beyond the scope of this text, as it is presented elsewhere (e.g., FAO, 1983). Here, current practices are illustrated for an arbitrarily selected case study to allow a more specific discussion and formulation of suggestions for future modifications. The case study, which is representative for problems elsewhere, involves grassland farming on heavy clay soils in the Netherlands (Bouma, 1981). These clay soils have been improved by tile drainage and fertilization practices. In the Dutch system of soil survey interpretation used in the 1960s and 1970s, these soils were indicated to have severe limitations (poor suitability) for modern grassland farming because of wetness in spring, poor trafficability in wet periods, and an inadequate moisture supply in drier years. A new system of land evaluation, based on FAO (1976), was introduced in 1979 (Haans, 1979). It defines three crucial land qualities as diagnostic criteria: moisture supply capacity, trafficability, and drainage status. Each land quality was defined in terms of soil properties that were either assembled during soil survey, could easily be measured in the field, or could be found in standard tables. So far, three major questions are key questions for any soil scientist applying the FAO framework:

1. What are the crucial diagnostic features to be considered when defining crop (or environmental) requirements and land properties?

Moisture supply capacity \ Drainage status / Trafficability	1+2		3		4			5	
	1	2	1	2	1	2	3	2	3
1	1.1	1.2	1.1	1.2	1.2	2.1	3.1	2.1	3.1
2	1.3	1.4	1.3	1.4	1.3	2.1	3.1	2.1	-
3	2.2	2.3	2.2	2.3	2.2	2.3	3.1	3.1	-
4	3.2	-	3.2	-	-	-	-	-	-
5	3.2	-	-	-	-	-	-	-	-

Figure 1. Simplified key to estimate land suitability for grassland in a modern, high-capital, low-labor, Land Utilization Type. Suitability classes 1.1, 1.2, and 1.3, suitable; classes 2.1, 2.2, and 2.3, moderately suitable; Classes 3.1, 3.2, and 3.3, unsuitable. The numbers for the three land qualities indicate different classes.

2. How are limitations or suitabilities determined?
3. How can diagnostic features be determined from an operational point of view?

Obviously, answers to these questions will determine the outcome of any evaluation. For grassland farming, the scheme was devised as shown in Figure 1.

The three selected land qualities were moisture supply capacity, drainage status, and trafficability. This selection is clearly a function of the particular land utilization type being considered, which is characterized by a high level of technology and relatively low labor requirements. The *moisture supply capacity* (as defined by Haans, 1979) was estimated in five classes by adding the volume of "available water" in the root zone to the volume of water flowing from the water table to the root zone by unsaturated flow during the growing season. Availability of a large volume of data on moisture retention (= pF curves) allows estimates of available water using standard tables that use soil texture as the diagnostic criterion. Estimation of upward unsaturated flow is a more complicated matter. Steady upward fluxes at different rates can be calculated for different soil materials from measured hydraulic conductivity curves using the water table level as a lower boundary condition and the negative pressure head at the bottom of the root zone as the upper boundary condition. Calculations can be made by computer, but also by hand. Calculations by hand allow easy estimation of upward fluxes in layered soils (Bouma, 1985). An upward flux of 2 mm day^{-1} was arbitrarily selected on the basis of the water deficit in the Netherlands in a dry year with a probability of occurrence of 10%.

The height to which a steady flux of 2 mm day^{-1} can be maintained is used as a measure for estimating whether or not an upward unsaturated flux can reach the root zone from the water table. The foregoing criteria provide a rather generalized estimate of the moisture supply capacity because many estimates are involved. The *drainage status*, as defined by Haans (1979), was defined in five classes by the average highest water-table levels which are routinely defined during Dutch soil survey (van der Sluys and de Gruyter, 1985). Drainage status, as defined here, is comparable to the aeration status defined elsewhere (e.g., FAO, 1976). *Trafficability* was defined in three classes in terms of the penetration resistance to be determined in the field with a portable penetrometer during the early part of the growing season when the water table is at its average highest level.

Data for the three land qualities just defined are derived from soil survey data and standard tables (moisture supply capacity), soil survey data (drainage status), and a simple field measurement (trafficability). This analysis answers the earlier question as to which diagnostic features are analyzed from an operational point of view. The two other questions relating to the particular choice of diagnostic features and the derivation of land suitabilities have yet to be discussed. Both are based on practical experience that indicates the importance of moisture supply in the growing season, when there is a water deficit, and excessive wetness and trafficability in the winter and early spring, when there is a water surplus. Soil fertility is not a problem in this land utilization type where nitrogen applications reach 300 kg per hectare. The suitability table presented in Figure 1 combines the three diagnostic land qualities into three suitability classes, each of which is subdivided into three subclasses. The heavy clay soil of the case study being discussed was classified in class 2.3, which indicates a restricted moisture supply in summer (100–150 mm) and a drainage status associated with an average highest water table level of 25 cm below surface. Obviously, this information is more detailed than a simple, generalized statement about a limited suitability (or moderate limitations).

The preceding discussion also demonstrates, however, the somewhat arbitrary character of the manner in which the various diagnostic features are defined and in which estimates of gradations are made. An alternative procedure is described next.

B. Quantitative Procedures

Two main problems are experienced when using the land evaluation procedure described in the previous section: (1) Separate diagnostic factors, which are often not independent, must be defined. The three land qualities that were distinguished are all direct functions of the soil water regime, and they are interrelated. Poor trafficability, for example, is associated with wet conditions, which also define drainage status and moisture supply capacity. (2) The need for operational procedures allowing use of soil survey data, standard tables, or simple field methods imposes a strong inflexibility on the system. The water supply capacity varies as a function of climatic conditions and fluxes of water in the soil. The field capacity and wilting point limit for available water represent a static measure for a highly dynamic natural phenomenon. The upward flux of water from the water table is handled crudely. Using water table levels only as a measure for drainage status does not allow for consideration of different natural drainage rates of surface soils of different textures and organic matter contents. Finally, trafficability is difficult to correlate with the resistance encountered when a cone

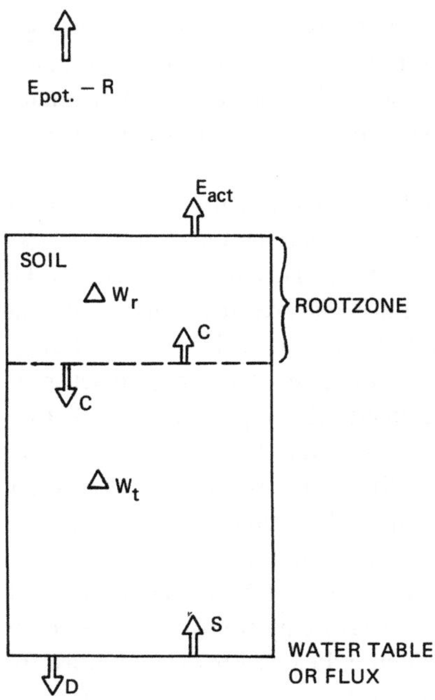

Figure 2. Schematic diagram showing fluxes calculated when simulating water movement in soil. E_{pot} potential evapotranspiration; R, precipitation; E_{act} actual evapotranspiration. *Arrows* reflect fluxes. W, changes in storage; C, flux at bottom root zone; D, downward; S, upward flux at water table.

is being pushed into the soil in the wet period of the year. Trafficability may also be a problem in other periods of the year, and it is really necessary to know its magnitude as a function of the moisture content or pressure head in surface soil in different times of the year.

Considering these problems, alternatives must be found. One successful alternative is presented by computer simulation of the soil water regime, which predicts water contents as a function of climate, water table fluctuations, and water extraction by crops. Simulation techniques have amply been discussed elsewhere (e.g., Feddes et al., 1978; van Keulen and Wolf, 1986; Wagenet and Hutson, 1987).

The basic principle (Figure 2) implies calculation of water fluxes in the soil as a function of environmental conditions imposed by the climate, water table fluctuations (if any), and crop growth requirements using hydraulic conductivity, moisture retention, and rooting data (Section IV). Dynamic simulation can be used to overcome the type of problems described, as is analyzed now for the various land qualities:

1. With simulation, the *moisture supply capacity* can be based on a more realistic assessment of water fluxes in the soil. The root zone is not considered to contain a fixed volume of "available" water, but a daily varying volume as a function of real rainfall, evapotranspiration, and varying vertical down or upward unsaturated fluxes moving out of and into the root zone. Calculations can be made for different years. Of particular interest is the possibility of making calculations for potential conditions, thereby exploring soil potentials.

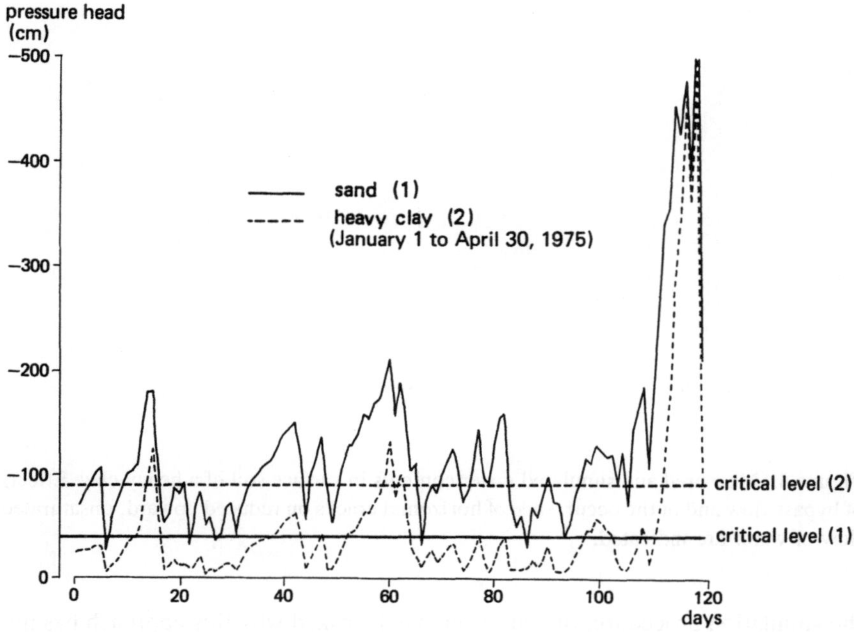

Figure 3. Plot of calculated pressure heads for a depth of 5 cm below surface in a heavy clay and a sand. These values are used to determine the number of days with adequate trafficability in any given period, using threshold levels of the pressure head, that were determined independently in both soils (from Bouma and van Lanen, 1987).

2. *Drainage status* can be calculated in terms of the air content in the root zone as a function of all factors mentioned under the previous point. Obviously, the drainage status will be influenced by the water table level (which is the only diagnostic factor being considered in the current evaluation system, as discussed), but other factors play an important role as well. Simulation allows daily expressions for any given period. reflecting different weather conditions. This is attractive because drainage status may be relevant for different periods in the growing season and not only for the moment when the highest water table level obtains.
3. *Trafficability* can be expressed as a function of the water content (or pressure head) of surface soil as calculated by the model. This requires definition of a "critical" water content, above which trafficability is inadequate (see Section IV). The number of days with adequate trafficability can then be presented as in Figure 3 for a sand and for the heavy clay being considered in our case study. Critical pressure heads (h) were −40 and −90 cm, respectively. The number of days with adequate trafficability are significantly different for both soils for the arbitrarily selected period indicated in Figure 3.

Simulation of the soil water regime occupies a central position when defining procedures for quantified land evaluation. Considering the obvious advantages of

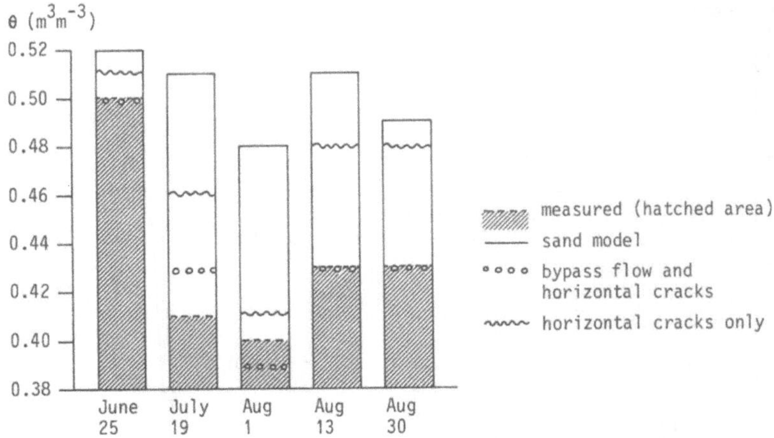

Figure 4. Measured and simulated water contents in surface soil of a heavy clay. Effects of bypass flow and of the occurrence of horizontal cracks on reduced upward, unsaturated flow of water are indicated.

the simulation procedure, the question may be asked why this approach has not yet been followed more widely. Three answers come to mind: (1) Simulation models have only recently become operational, following satisfactory field testing of modeling results. (2) Basic data for the models were not available. Rather than use hypothetical quantitative data in detailed models, it is indeed more realistic to focus on qualitative interpretations that are more in agreement with limited available data. (3) Even when basic data such as hydraulic conductivity and moisture retention characteristics were available, they produced quite unrealistic results when used as input data in simulation models. The reason was the often heterogeneous character of field soils, which deviates strongly from the underlying assumptions of flow theory that assume the presence of homogeneous and isotropic porous media flow (Klute, 1973).

Practical use of simulation techniques for quantified land evaluation and its problems will finally be illustrated for the case study of heavy clay soil being considered here. Because using simulation to characterize the soil water regime seemed relevant, measurements were made of hydraulic conductivity and moisture retention using modern standard techniques (Bouma and de Laat, 1981). Running a simulation model on real-time basis for the summer of 1979 produced absurd results (Figure 4). Field studies indicated the importance of bypass flow (vertical movement of free water along air-filled macropores in an unsaturated soil matrix) and the effects of horizontal cracks on the upward flow of water from the water table to the root zone. New field techniques were devised to take these phenomena into account. Some of these are discussed in Section IV; detailed descriptions have been presented elsewhere (e.g., Burke et al., 1986).

The role of soil survey specialists should go beyond definition of representative soils and soil horizons in which *others* make measurements of physical and chemical characteristics. They should be involved when defining these measurements and considering site conditions as well as when interpreting results of calculations.

III. Modern Land Use Questions

A. Agriculture Versus Environment

Mankind is dependent on the soil either directly or indirectly for food, clothing, and shelter. Whatever technological advances may occur in the future, the soil must continue to produce the food necessary for feeding the ever-increasing population of the world. The greater part of mankind still knows the grim meaning of hunger. Many of the minority that do not now associate food with the supermarket, rather than with nature, take the continuing supply of food for granted rather than holding it as a gift to be cherished. It is forgotten that once we were all farmers in an agricultural world, a world that should not be idealized: lives were simple, but vulnerable to the whims of nature. The technological advance of the last century has raised the quality of the lives of many, but the cycle of life cannot be widened and diversified without penalties. We now face a basic challenge, which can be met by balancing our considerable technological skills with the demands of nature. This can be done without retreating to an idealized but unrealistic past in which each inhabitant of the world was living from his own land. Entire civilizations have disappeared because they overextended their demands on nature. Our challenge is to avoid this and to define a way of life in harmony with nature, allowing agricultural production to be sustained.

Environmental concerns are increasingly reflected in environmental laws limiting chemical fertilization and application of pesticides and herbicides. One of the major challenges for soil science in the years ahead is therefore to help define optimal application levels of agricultural chemicals to develop corresponding operational management schemes. If soil scientists do not face up to this challenge, others will. Much work is in progress on movement of chemicals in soils. However, many studies focus on specific disciplinary aspects that are usually studied in considerable detail. There is a need to also focus on broader studies that consider areas of land with its inherent spatial variability patterns and, particularly, on the interaction of physical, chemical and biological processes. Land evaluation, as defined by FAO (1976), has the pretension to analyze land suitablities in a social and economic context. More than ever, this approach is necessary in a society in which food surpluses are becoming structural in character in many countries and the environmental side effects of agriculture are bound to receive ever greater emphasis.

Studies of nutrient regimes on farm level in various countries demonstrate the feasibility to reduce application rates of chemicals without reducing crop yields. This development is particularly attractive for farmers, as it holds the promise of reduction of costs. Such management systems can only be developed, however, if soil studies are multidisciplinary in character. Not only water fluxes are important, but also the associated nutrient transport, which is partly a function of crop uptake of nutrients and microbiological transformations in the soil.

B. Problem Analysis

Soil survey interpretation and land evaluation procedures often follow rather rigidly defined procedures. This is necessary because they have to be applied in many soil survey studies. One possible disadvantage of such uniform standard procedures is the implied impression that no problems remain to be solved. This, of course, is not true. It is of utmost importance to be constantly aware of new questions being asked and of the particular interests of those asking the questions, and to think of ways in which soil survey data can be used to answer the questions, sometimes perhaps only partially. Too many standard answers appear to be desperately in search of the questions that led to their formulation some time ago. Soil survey has, by tradition, provided a particular answer for questions with differing degrees of detail by making studies at different scales. Some questions arise, for example, from planning activities at a regional scale, where only generalized statements can be made about soil properties for relative large areas of land so as to be in agreement with the occurring spatial variability. Other questions are quite detailed, for example, when directed at farm level, and they need more specific answers.

Providing specific answers in quantitative terms is often very difficult because of lack of data. This presents a basic and serious dilemma for those who have to make statements about properties of areas of land. Two aspects of this problem are discussed in the remainder of this chapter: (1) The character of soil data available for land evaluation and its relative areal applicability, and (2) quantitative land evaluation at different scales, including a tentative assessment of accuracy obtained.

The diagram in Figure 5 (Bouma, 1986) illustrates the way in which the important land quality: "moisture availability" can be obtained. Originally, land evaluation consisted of gathering the farmers' experience and the opinions of experts. Results obtained covered the area being considered. More quantitative approaches, for example, those calculating "available water," required specific moisture retention data ($h - \theta$ relationship) that were not available at first: Areal applicability was reduced until this deficiency was corrected. A further reduction occurs with the introduction of simulation models. Simple models, such as the de Laat (1980) model, make calculations for 7-day periods and consider calculated potential evapotranspiration values. More complex models, such as those presented by Belmans et al. (1983), require more specific climatic and soil data, while calculations are made for shorter time periods. This makes the model potentially more accurate, but also more costly and complicated. Finally, detailed simulation of, for example, basic transport phenomena such as water movement from the soil into root hairs is very relevant from a scientific point of view but has little applicability, as such, for land evaluation studies that cover large areas of land. The researcher in land evaluation must choose his appropriate

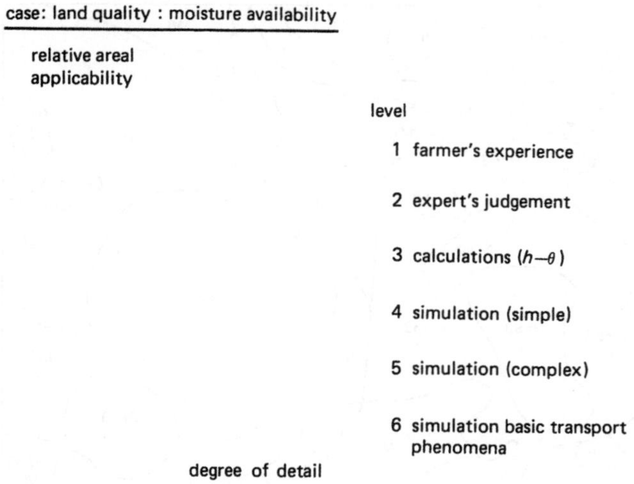

Figure 5. Schematic diagram illustrating different levels of detail in research as compared with the relative areal applicability in land evaluation (after Bouma, 1986).

level of generalization that, on the one hand, is sufficiently detailed to allow quantitative and reproducible calculations but, on the other hand, can still be applied in an operational context for areas of land. To allow the application of the attractive level 4 (Figure 5), attention must be paid to use of pedotransfer functions, as discussed in Section IV.B (Bouma and van Lanen, 1986).

The diagram in Figure 5 is not complete without consideration of the type of questions being asked. Application of knowledge at levels 1 and 2 can produce excellent results at reconnaissance level. In fact, it may be quite undesirable to move too rapidly to quantitative simulation procedures if questions have a broader character than can realistically be characterized by simulation at this time. Specific questions, for example, covering moisture availability for a given crop in different years including the associated soil moisture regime, require a level 5 approach. The choice is then to focus on one particular subarea of land where detailed data are to be obtained rather than on inadequate procedures for a larger study area. The key aspect expressed by Figure 5 is the need for:

1. A critical analysis of the problem to be studied, requiring intensive contacts between the soil scientist and representatives of agencies asking questions
2. Insight on the part of the soil scientist as to what can be achieved with different types of available data and different levels of funding considering well defined questions.

This second point will now be illustrated, using a case study by Wösten et al. (1987). In the Netherlands, soil maps are increasingly used for environmental interpretations in the context of new laws on soil protection. Such interpretations have to be quantitative so as to allow reliable estimates of soil properties in rela-

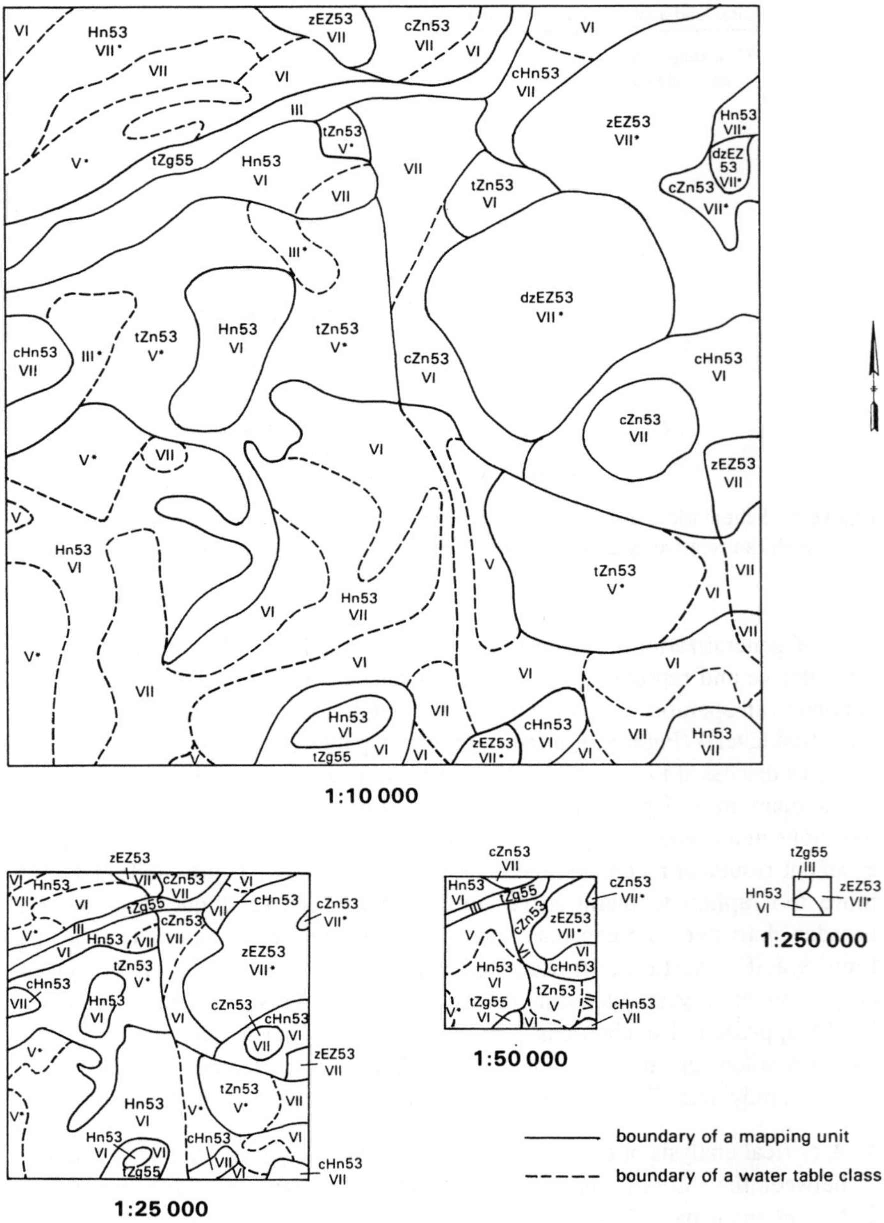

Figure 6. Soil maps of an area of 404 ha with sandy soils in the Netherlands, expressed in four scales (Wösten et al., 1987).

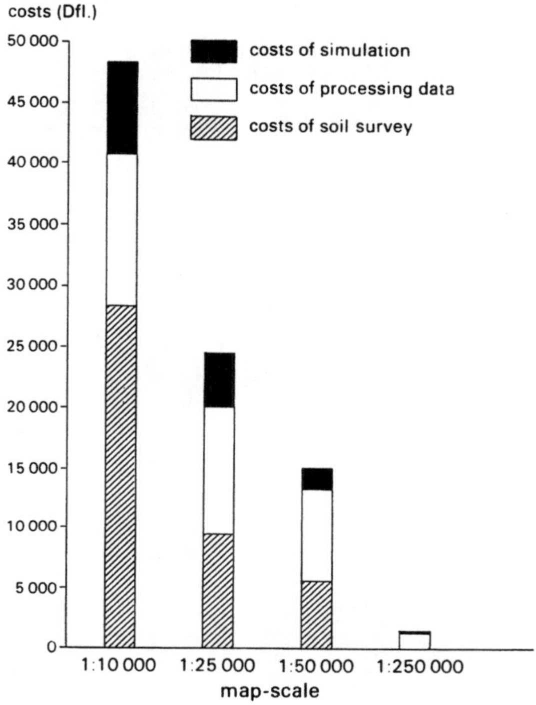

Figure 7. Relation between costs and map scale of maps shown in Figure 6 (Wösten et al., 1987).

tion to critical levels that are being defined in soil protection laws. The case study to be discussed deals with the effects of lowering of water table levels on grass production. Lowering of the water table results from water extraction for municipal water supply. Farmers have to be financially compensated for damages incurred, and some of them take claims to court.

Soil maps of an area with sandy soils of 404 ha were made at four different scales (Figure 6). The largest scale map (1:10,000) shows most details in terms of soil types, as expressed by the codes on the map, and water table fluctuations as expressed in classes indicated by Roman numerals. For details, the reader is referred to Wösten et al. (1987). The number of delineated areas on the maps decreases with decreasing map scale. Maps are made by soil surveyors who translate point observations, made using an auger, into areal patterns by observing various land features such as slope, vegetation, etc. The number of observations decreases as the scale of the map to be published decreases. Here, 606, 135, 54, and 1 borings, respectively, were made for the four maps. This is associated with a sharp reduction in costs as the map scale decreases (Figure 7).

Traditional soil survey interpretations define a "representative" soil profile for each mapping unit, based on the experience of the surveyor who synthesizes all his observations. Maps were used to derive basic soil physical data

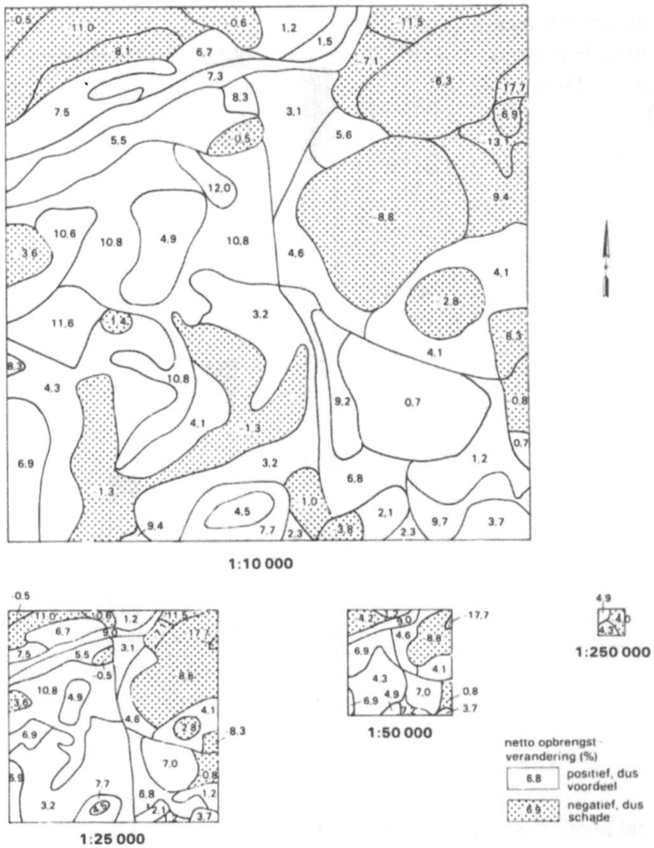

Figure 8. Segments of maps of the sample area with changes in yield calculated on four scales. In each case, a "representative soil profile" was defined for each mapping unit (Wösten et al., 1987).

for the various "representative soils," while each map also indicates the areal extent of each soil type. Computer simulation was used to obtain quantitative and reproducible predictions (Bouma et al., 1980). Soil maps (Figure 6) were used to make calculations for each unit, on the basis of a defined "representative" soil profile. This involved application of different procedures (see Section IV). Results were presented in terms of the absolute change of grass production, comparing the "old" situation *without* and the "new" situation *with* groundwater extraction. One percent change corresponds with Dfl. 50.—. Results obtained are shown in Figure 8. Two questions were being asked: (1) What is the average change in yield per year for the entire area? (2) What is the change in yield per year for any given parcel of land? Average changes per year, considering all delineated areas of the various maps, were calculated. They were 2.7%, 2.5%, 2.6%, and 2.5% for maps on scales 1:10,000, 1:25,000, 1:50,000, and 1:250,000, respectively. In other words, the relatively cheap small-scale maps

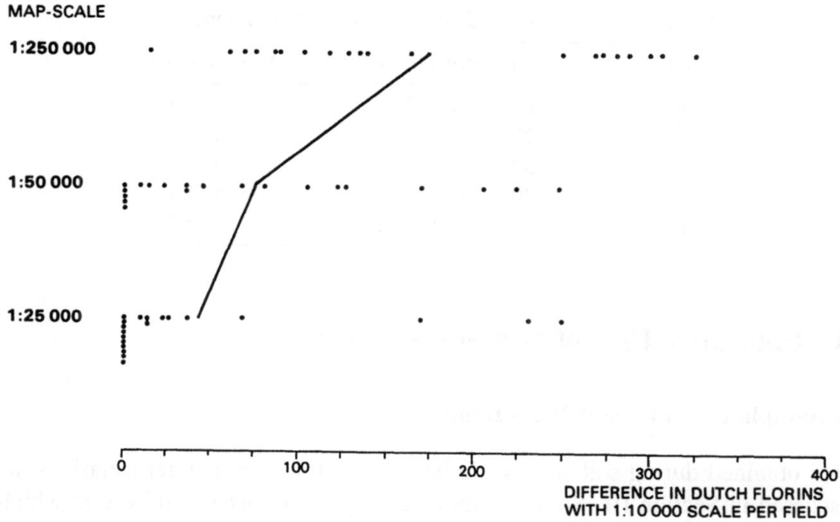

Figure 9. Absolute differences in yield, expressed in Dutch florins per parcel per year, of 20 randomly selected parcels calculated on four map scales with the 1:10,000 map as a reference. Average absolute differences in yields are indicated by line connecting averages for each scale.

are quite suitable to estimate the average change in yield for the entire area. To analyze the second question, 20 parcels were selected at random and they were located on the different maps. Calculations were made for the soil types occurring within the parcels. Results are presented in terms of deviations from results of the 1:10,000 map, which is considered as a reference (Figure 9). This is acceptable because the cost of larger scale surveys would be prohibitive.

Deviations increase as the scale of the map decreases. The average deviation per parcel per year is Dfl. 43, Dfl. 76, and Dfl. 173 for the 1:25,000, 1:50,000 and 1:250,000 maps, respectively. More significant, perhaps, is the observation that deviations for individual parcels can be very high, up to Dfl. 300 (Figure 9). On the other hand, no differences were observed for 10 parcels at the 1:25,000 scale and for 5 parcels at the 1:50,000 scale. Each landowner must balance the cost of a survey versus the benefits resulting from the calculations. The conclusion here was that the higher cost of a 1:10,000 survey was less important than the associated increased accuracy of the calculation. Using cheaper smaller scale surveys resulted in more variable data, which would be less valuable when making a claim in court. This illustrates what might well become a standard phrase in soil survey: "You pay for what you get." The example is meant to illustrate that the scale to be chosen for field work in soil survey should not primarily be a function of soil variability but also of the type of question being asked and of a specific cost/benefit analysis. This aspect has largely been ignored within soil survey so far with some notable exceptions (e.g., Bie and Ulph, 1972). However, emphasis has generally been on countrywide surveys at a given scale using a fixed legend.

Table 1. Dimensions and volumes of in situ columns

Column	Dimensions (cm)	Volume (cm^3)
A	160 × 75 × 20	240,000
B	120 × 75 × 20	120,000
C	50 × 50 × 20	50,000
D	20 (diam.) × 20	6,283
E	7 (diam.) × 6	884

IV. Innovative Uses of Soil Survey Data

A. Sampling and In Situ Measurements

Data obtained during soil survey are being used for various interpretations, as discussed in Section III. Some arbitrary examples are discussed here in which such data are used to estimate optimal sample volumes and representative placement of monitoring equipment.

The *first* example is concerned with defining optimal sample volumes in soils with natural aggregates ("peds") and continuous large soil pores, such as cracks and root and worm channels. When defining physical methods for measuring hydraulic conductivity, handbooks usually emphasize technical and theoretical aspects, while considerably less attention is paid to operational aspects.

Discussions on spatial variability of soil hydraulic characteristics focus on mathematical manipulation of measured data, which are usually accepted as presented. However, part of the observed variability may result from using and incorrect method, or an unrepresentative sample volume, particularly in soils with peds, Anderson and Bouma (1973) showed that the saturated hydraulic conductivity (K_{sat}) of a silt loam B2t horizon, as measured in soil cores, varied considerably as a function of core height. Average values ranged between 650 cm day^{-1} for 5-cm-high cores to 100 cm day^{-1} for 17-cm-high cores. The latter had a volume of 270 cm^3, and contained approximately 20–30 blocky peds. Anderson and Bouma explained the observed differences by considering vertical continuity patterns of planar voids ("cracks") between peds. The hydraulic conductivity of a sample is determined by the smallest crack in the flow system, which forms a "neck." The probability of occurrence of a small neck in a vertical flow system increases as this system becomes longer. Beyond a certain length, "necks" always occur, and this results in a decreasing effect of sample length on measured K values.

Comparable results were reported by Lauren et al. (1987) when measuring infiltration rates into the B2t horizon of a Hudson silty clay loam. They compared rates obtained in plots of five sizes (Table 1). Results (Table 2) indicate higher rates and standard deviations for the smaller samples. Column A showed irregular rates because of operational problems associated with ponding of a very large horizontal area of 12,000 cm^2. The optimal volume of column C corresponded

Table 2. Statistical parameters for K_s measured by different sampling volumes

Size	Mean[a]	Mode[a]	Median[a]	SD[a]	CV (%)	Number of samples
A	21.3	10.3	16.6	16.9	79	37
B	13.7	6.4	10.7	11.0	81	36
C	14.4	6.3	10.9	12.5	96	37
D	36.6	6.3	20.3	54.9	150	37
E	34.5	4.8	16.3	64.0	186	35

[a] Values in centimeters per day.

with approximately 60 peds. Perhaps a somewhat lower number of peds could still be acceptable, considering relatively large differences between the volumes of columns C and D (Table 1). Be that as it may, these examples illustrate relationships between optimal sample volumes and ped volumes, the latter to be estimated from soil structure descriptions. Such relationships should be further explored for other soils and structures to allow reliable field estimates of optimal, representative sample volumes. These are bound to be different for different soils, when measuring hydraulic conductivity.

The *second* example is use of unlined auger holes or piezometers in soils to measure water table levels. Open auger holes are commonly used to measure water table levels (Soil Survey Staff, 1975). Problems may occur in soils with peds where water may flow into the auger hole through its sidewalls, which may intercept continuous macropores connected to the soil surface (Figure 10). Bouma et al. (1980) demonstrated very rapid movement of water, containing chlorides as a tracer, from the surface to the water table at 80 cm depth in a wet, heavy clay soil. This water was intercepted in open auger holes at different depths above the real water table, incorrectly suggesting the presence of perched water tables. Use of piezometers with closed sidewalls did not allow this flow into the auger holes, thus indicating unsaturated conditions that did indeed occur as confirmed by tensiometer measurements. Consideration of flow patterns in a clay soil can help to explain what appear to be quite erratic measurements of water table levels when using standard techniques. Reliable and less variable data are obtained when measurements are modified, taking into account particular morphological features of the flow system. One aspect, as illustrated in Figure 10, illustrates the undefined character of water table levels in soils with macropores. Water will preferentially move through the macropores and will fill macropores above the water table until the water is laterally and vertically absorbed by the soil, leaving a horizontal plane of zero water pressure. For a certain period after rainfall, therefore, water tables, as traditionally defined, do not exist in soils with macropores. It is important to know the duration of such periods to avoid erratic measurements.

The *third* example deals with the measurement of infiltration rates into soils with continuous macropores, be it worm channels or cracks. Traditional infiltra-

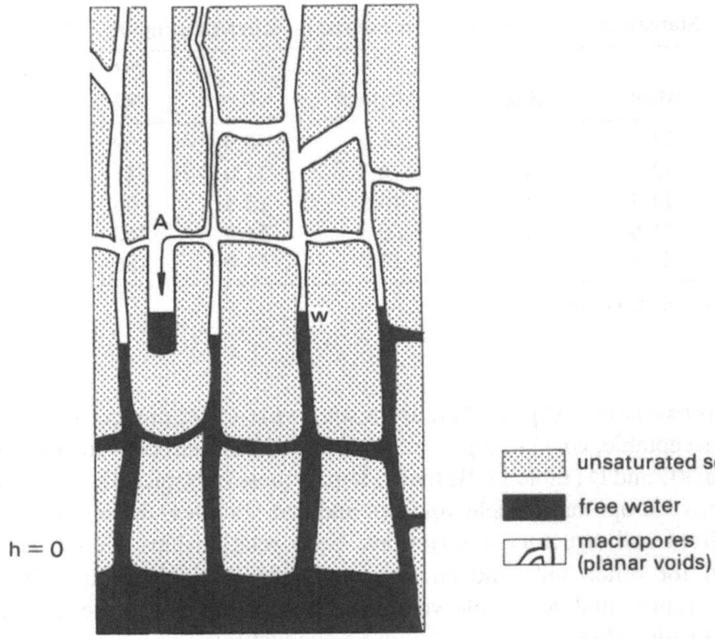

Figure 10. Schematic diagram illustrating undefined character of water table after rainfall or irrigation when free water occurs in planar voids (*w*) surrounded by unsaturated soil. Occurrence of free water in an unlined auger hole (*A*), as illustrated, does not indicate presence of a water table.

tion experiments assume the presence of homogeneous soil in which the infiltration front will penetrate as a more or less regular plane. Thus, measurement of one flux density is realistic. In soils with macropores, rapid movement occurs through the macropores while water inside the soil matrix is more or less stagnant. Two procedures can be followed: (1) A count is made of the number of channels (or cracks) in horizontal cross section, and the horizontal infiltrative area is chosen sufficiently large to include a representative number of channels or cracks. (2) Infiltration rates are measured separately into macropores and into the matrix. Measurement in individual channels were reported by Bouma et al. (1982) and in cracks by Bouma and Wösten (1984). Infiltration rates into the entire soil are computed by counting the number of channels and the length of cracks, and by adding the volume of infiltration of water in macropores to the volume that infiltrates vertically into the soil matrix between the macropores. An infiltration rate is obtained by dividing the total volume of water by the total area of infiltration. This procedure allows a clear understanding of the effects of macropores on infiltration, because a series of infiltration rates into individual channels is obtained.

A striking difference is observed when comparing infiltration rates in soil as measured in cores that are attached to the subsoil, with K_{sat} values in detached

Table 3. Statistics for K_s measured on detached and attached columns

Method	Mean	K_s Median[a]	Mode[a]	SD (cm day^{-1})	CV (%)
Attached	36.6	20.3	6.3	54.9	150
Detached	4,296.6	286.6	1.3	64,280.2	1,496

[a] K_s in centimeters per day.

cores taken from the same soil. Differences (as shown in Table 3, from Lauren et al., 1987) result from continuity patterns of the macropores. Flow in the soil matrix occurs in both cases at unit gradient of the hydraulic head. Theoretically, the infiltration rate is then equal to K_{sat}. Obviously, occurrence of macropores presents a problem: when continuous in detached cores, they conduct much water, but when discontinuous in attached cores, they just fill with water.

Definition of unsaturated soil conditions becomes difficult as well. Application of a series of mechanical crusts with increasing hydraulic resistance resulted in a series of decreasing infiltration rates which all occurred at zero pressure and unit hydraulic head gradient, as measured by tensiometers (Bouma, 1982). Thus, all these infiltration rates represented K_{sat} values according to the common definition. Saturation was not associated with one K_{sat} value but with a range of values, for example, from 2000 cm^{-1} to 10 cm day^{-1} for a heavy clay. Once the crust resistance exceeds a critical value, unsaturated conditions are induced in the underlying soil, and this, obviously, represents a classical unsaturated flow system. Differences observed by Bouma (1982), as described here, were attributed to decreasing flow of water along the walls of air-filled macropores. Such processes are common in the field, and their recognition helps to explain what may appear to be erratic and highly variable observations to which classical concepts of water flow are assumed to apply.

The foregoing example illustrates the presence of elements of variability that are introduced in soil hydrology when macropores occur in the flow system. Knowledge about the type of pores and their abundance allows a better characterization of the flow process. Quantitative soil structure descriptions have a useful function here.

B. Obtaining Basic Data Simulation Modeling

Simulation models form a crucial element of quantitative land evaluation. A central problem when using these models is lack of suitable data. Climatological data are often available, but may be in a generalized form that is inadequate for specific calculations. Soil data, such as hydraulic conductivity [$K(\theta)$] and moisture retention [$h(\theta)$], are often lacking. The same is true for information on rooting patterns and water table fluctuations. The simulation modelers are faced with a very basic problem here. They can proceed by making assumptions that appear to be reasonable. Having chosen that avenue, they are usually forced to further proceed by "tuning" and "matching" calculation results when these deviate substantially from measured values. This "matching" has no scientific basis, and a reader of reports presenting results of simulation runs should critically examine the usual types of graphs in which calculated and measured points always appear to coincide in a gratifying, satisfactory manner. The basic strength of simulation modeling is its ability to divide a highly complex reality into interrelated subprocesses that are mathematically defined. Thus, relationships among many interacting processes are made visible, producing a welcome degree of clarity and transparency in what used to be an amorphous and rather overwhelming accumulation of variable data.

Returning to use of soil survey data, it is clear that many basic soil data are available. Most soil survey organizations have geographical information systems, and developments in this field are very rapid. Data being stored are of the type that are being gathered in the field, for example, classification, soil horizons, color patterns, texture, structure, contents of organic matter, carbonates, etc. Most of these data are not directly applicable for simulation. A major challenge for soil science is to "translate" data we *have* to data we *need*, if only because there will not be funds available to obtain, for example, $k(\theta)$ and $h(\theta)$ data on a large scale. Procedures to obtain these "translations" have been discussed by Lamp and Kneib (1981), who defined pedofunctions, and Bouma and van Lanen (1987), who used the term transfer functions. To avoid confusion with the same term being used in soil physics (Jury, 1982; Jury et al., 1982), the term pedotransfer functions will be used hereafter. Pedotransfer functions relate different soil characteristics and properties with one another or to land qualities. The terminology being used here is based on FAO (1976), as modified by Bouma and van Lanen (1987). Two types of pedotransfer functions are distinguished: those that are continuous (e.g., using percent clay, percent organic matter, etc.) and those that relate to distinct classes (using a horizon designation, a soil type, etc).

An example of using pedotransfer functions is shown in Figure 11 for the land quality soil-water-deficit. Six soil characteristics are listed that can directly be derived from soil survey reports. Four of those are continuous characteristics; soil structure and horizon code are class characteristics. Two land characteristics cover the legend unit of the map and the groundwater class (Gt), which defines the mean highest and mean lowest level of the water table (van der Sluijs and de Gruijter, 1985).

Simulation of the soil water regime requires hydraulic conductivity and moisture retention data, water table levels (Gt), rooting depth, and climatic data. These requirements have been widely discussed elsewhere (Feddes et al., 1978; van Keulen and Wolff, 1986). The model will calculate the real evapotranspiration, which is often lower than the potential evapotranspiration, during the average Dutch growing season. The difference is the water deficit, which can be calculated on a daily basis, for single years or for longer periods of time. Figure 11 contains several pedotransfer functions (TFl) that are now discussed. A continuous pedotransfer function (TF1) has been derived by regression analysis to relate soil bulk density (b) to texture and organic matter contents. As an example:

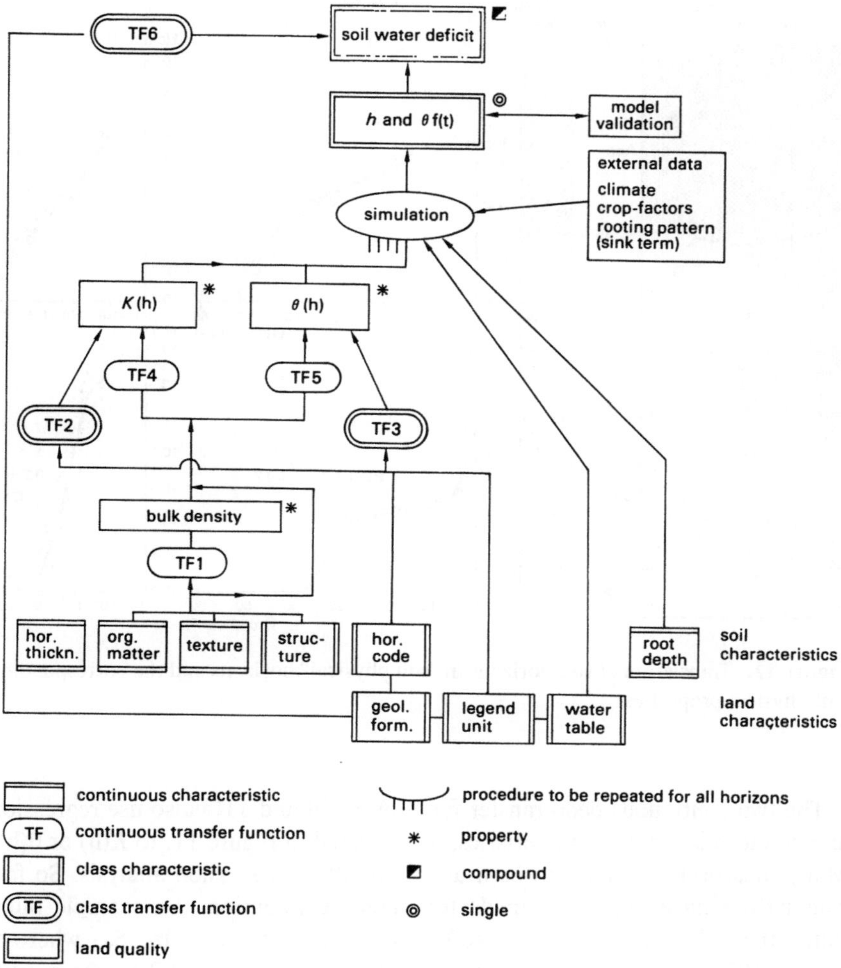

Figure 11. Flow diagram for the land quality Soil Water Deficit (Bouma and van der Lanen, 1987).

$b = b + b_1H + b_3M$, where b, b_1, b_2, and b_3 are characteristics constants, H is organic matter, S is percent and silt, M is median sand fraction.

Standard deviations are relatively high. The decision whether to measure a characteristic or to calculate it will depend on an analysis of the amount of effort it takes to make a measurement in relation to the associated improvement of results obtained. Measurement of bulk density is simple and involves little cost. Use of TFl is therefore likely to be limited. Important pedotransfer functions are distinguished to derive the soil properties hydraulic conductivity [*K*(h)] and moisture retention [θ(h)]. Both measurements are complex, time consuming, and costly, requiring considerable expertise.

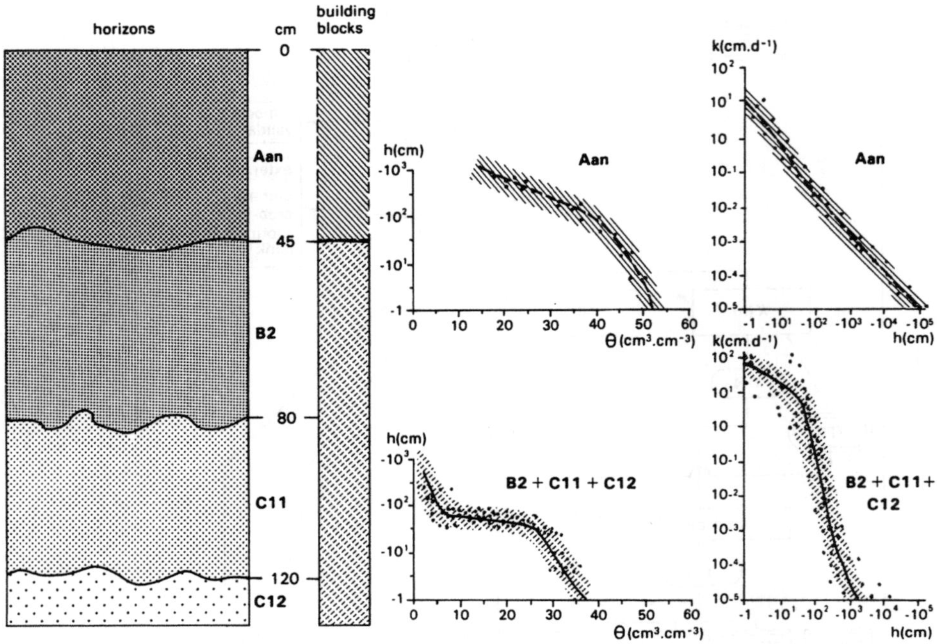

Figure 12. Translation of soil horizons in "soil physical" horizons and the corresponding soil physical properties.

The two continuous pedotransfer functions TF4 and TH5 also use regression analysis to relate soil characteristics, as indicated in Figure 11, to *K*(h) or θ(h). Many measurements have to be available to allow regression analysis. So far, calculations have only been made for moisture retention. An example illustrates the principle: $\theta\ (h = -100\ \text{cm}) = b_0 + b_1.L + b_3.\ S_y$, where θ ($h = -100$ cm) is the water content at a pressure head of -100 cm (formerly: pF 2.0), b_0 through b_4 are coefficients determined by regression analysis, H is percent organic matter. L = percent clay, and S_y is 1/dry bulk density. The procedure outlined differs from other procedures that use textural data to calculate *K*(h) and θ(h) values directly or procedures that use measured moisture retention data to calculate *K* (Green and Corey, 1971). A detailed discussion of these procedures relies on a "sand" model of capillary tubes to "translate" static soil characteristics, such as texture and moisture retention data, into the dynamic hydraulic conductivity. Poor results are obtained in structured soils containing clay and silt. Returning to the procedure by which *measured* properties are related to characteristics, a need can be distinguished to characterize conductivity and retention *curves* as a whole, rather than in terms of separate points, as illustrated above for moisture retention. The procedure of van Genuchten (1980) can be applied in this context.

The two class pedotransfer functions TF2 and TF3 both relate physical soil properties to well defined horizon designations for soil series as distinguished in soil survey. Soil horizons have a geographical distribution pattern, as indicated on the soil map for the various soil series in which they occur. Thus, the basic problem of extrapolation of point to area data is solved, by using pedogenic soil horizons as carriers of information. The discussed approach was earlier presented by Baker and Bouma (1975) and Baker (1978). A recent example was presented by Wösten et al. (1985), who characterized physical soil properties of 12 different pedogenic soil horizons, in an area of 800 ha, to find that only 6 of those were significantly different from a physical point of view. This illustrated the important phenomenon that pedogenic differences do not necessarily correspond with functional differences, such as differences in physical properties (Figure 12). McKeague et al. (1984) related soil structure to K_{sat}, which represents a comparable procedure. A similar conclusion was reached for environmental land qualities by Breeuwsma et al. (1986). Breeuwsma et al. related the phosphate sorption capacity (PSC) to the oxalate extractable Fe and Al by $PSC = 0.4\ (Al_{ox} + Fe_{ox})$. Thus, the cumbersome direct measurement of the PSC could be replaced by a procedure using available data assembled in the context of pedology studies.

Breeuwsma et al. related the cation-exchange capacity (CEC) to the organic matter (OM) and clay content by $CEC = 1.5\ OM + 0.5\ clay$. Finally, travel times were derived from soil physical data for soil horizons as described for TF2 and TF3. Applying these pedotransfer functions to "representative" profiles of each mapping unit allows derivation of maps that express aspects of environmental vulnerability, as shown in Figure 13 for the PSC.

The examples of Breeuwsma et al. (1986) again demonstrate that pedogenetic differences, as expressed by horizon designations, do not necessarily correspond with functional differences. Some different pedological horizons function in a comparable manner. However, different groups of horizons obtain for different functions, as is illustrated in Figure 14 for the examples discussed here, using a Plaggept soil as an arbitrary example. The eight pedogenic horizons are all used to estimate CEC because their organic matter contents are considered to be significantly different. However, for PSC only five, and for travel times only four, groups of horizons are different. Many other functional characterizations can be derived, for example, nitrification and denitrification potential based on organic matter content, mechanical stability based on texture and structure, etc. There remains as yet a large research field to be explored.

A rapid method to determine land qualities is to relate them to soil series. This is done by TF6 (Figure 11), which defines soil water deficit as a function of soil classification which includes soil series and water table fluctuation class. TF6 assumes that several calculations have been made for a particular soil series, allowing predictions for identical soil occurring elsewhere. For soil water deficit there may be a problem in terms of varying subsoil textures within a given soil series. These various textures may give rise to different rates of upward unsaturated flow yielding a range of deficit values. Functions such as VF6 relating to soil water regimes are therefore only feasible for soil series in which subsoil textures are identical. However, they allow the rapid development of derived maps from soil maps using a database containing data for soil series.

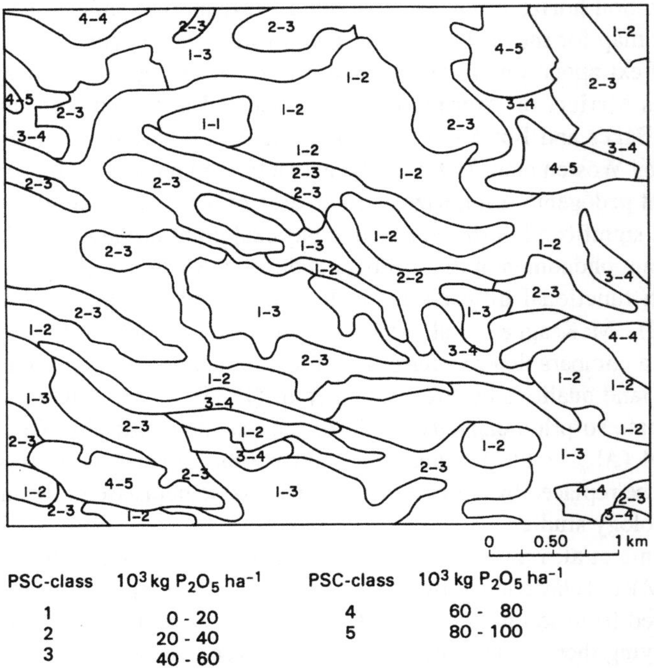

Figure 13. Map of the PSC classes of the unsaturated zone as a function of the mean highest water table (MHW, first number) and the mean lowest water table (MLW, second number). PSC, Phosphate Sorption Capacity (P = 0.44 × P_2O_5); MHW = Mean Highest Watertable; MLW, Mean Lowest Watertable (Breeuwsma et al., 1986). Reproduced from *Soil Science Society of America Journal*, Volume 50, No. 1, January–February 1986, pages 186–190, by permission of the Soil Science Society of America, Inc., Madison, Wisconsin.

The land quality soil water deficit is a function of water fluxes in the soil as such. Other land qualities are characterized by the consideration of threshold values that separate satisfactory from unsatisfactory levels. In this context, Figure 15 covers the land quality trafficability that is also a function of the soil water regime. In addition, a threshold value for the pressure head or the water content of surface soil is needed. When the soil is wetter than this critical value, trafficability is considered to be inadequate. When drier, trafficability is adequate. Similar considerations apply for the land quality aeration status of the root zone, for example.

For trafficability two new class transfer functions are defined as TF7 and TF8, which relate the threshold value to land and soil characteristics, respectively. Procedures to derive threshold values are not discussed here (see Bouma and van Lanen, 1987). Simulation of soil water regimes for several years and consideration of threshold values allow derivation of expressions for the probability of occurrence of adequate trafficability at any given time of the year. Figure 16

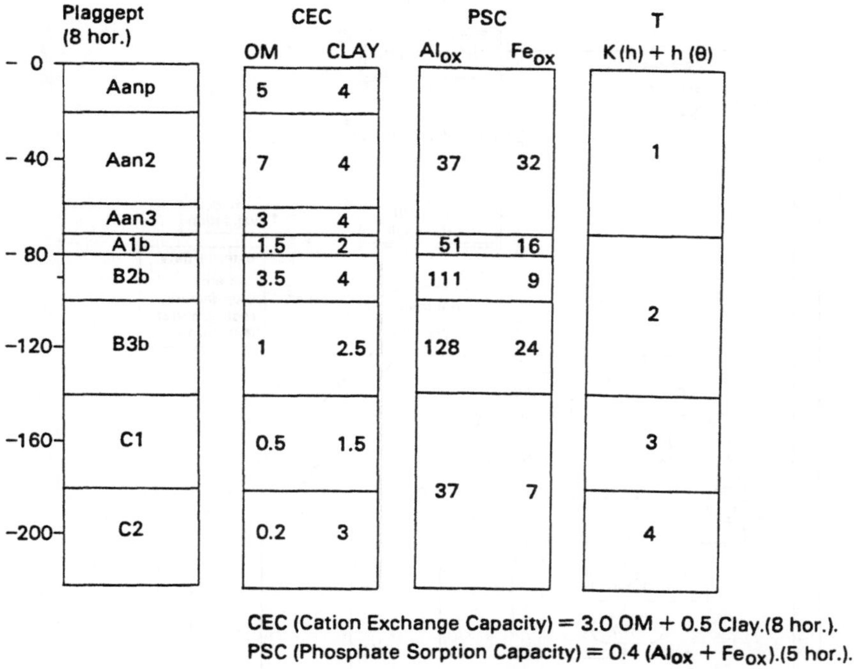

Figure 14. A Plaggept with eight soil horizons being interpreted for CEC, phosphate sorption capacity, and hydraulic characteristics to be used for calculating travel times. Eight, five, and four horizons behave differently, respectively, for the three types of interpretations being considered. CEC, (mmol ha^{-1}); PSC, 10^3 kg P_2O_5 ha^{-1}; T, days; K(h), m day^{-1}; h (θ), m.

shows the probability of adequate trafficability for a heavy clay soil, as based on simulation runs for a period of 30 years (Wösten and Bouma, 1985).

C. Modeling Water Regimes in Soils with Macropores

The role of soil structure descriptions when defining sampling and monitoring procedures was discussed by means of some examples in Section IA. Availability of representative data for $K(\theta)$ and $h(\theta)$ does not as such guarantee that water regimes in soils with macropores can be simulated accurately. White (1985) has written an excellent review of the state of the art. Beven and Germann (1982) have summarized a large body of experimental evidence indicating that infiltration and redistribution of water in soils with macropores are not adequately described by theories that treat the soil as a homogeneous medium conforming to Darcian principles of water flow. A further discussion of this topic in the context of this paper is relevant only inasmuch as it relates to use of soil survey information.

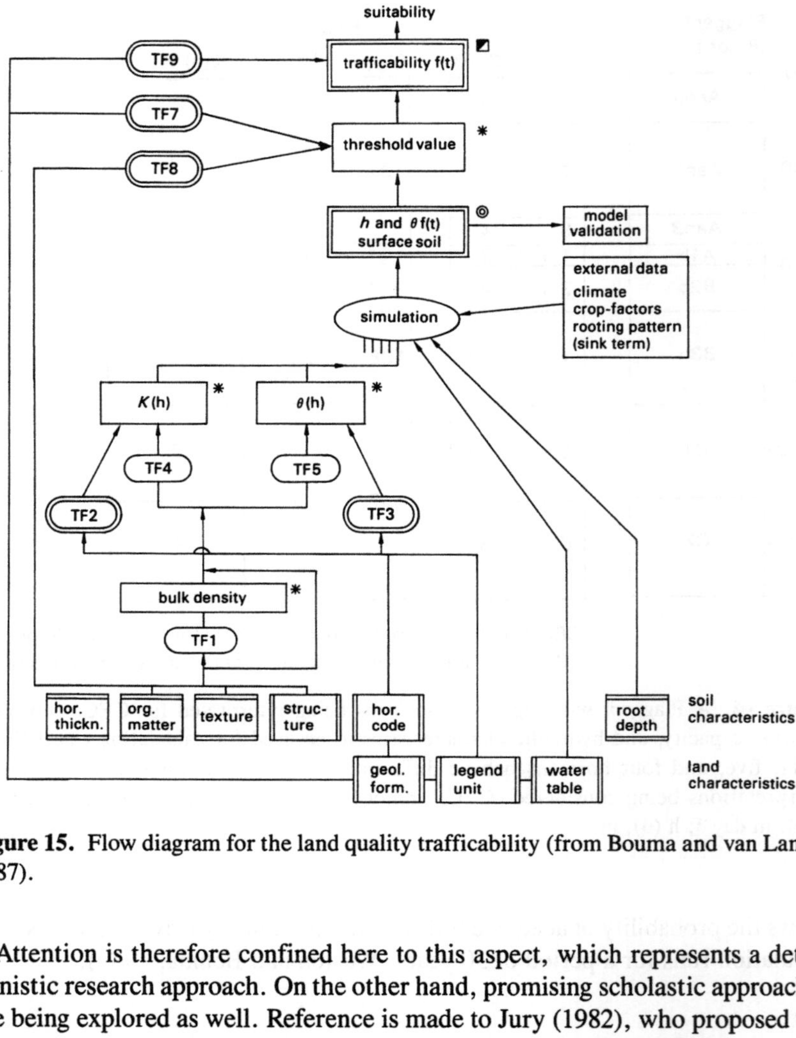

Figure 15. Flow diagram for the land quality trafficability (from Bouma and van Lanen, 1987).

Attention is therefore confined here to this aspect, which represents a deterministic research approach. On the other hand, promising scholastic approaches are being explored as well. Reference is made to Jury (1982), who proposed use of a probability density function, and to the review by White (1985).

Two approaches can be distinguished when simulating water movement in soils with macropores:

1. A detailed approach in which infiltration and redistribution of separate showers are analyzed. Vertical flow into macropores is assumed to occur 'after some surface ponding' when the infiltration rate into the soil matrix becomes lower than the rate by which water is applied to the soil surface. Infiltration of water into vertically continuous macropores at the soil surface is followed by lateral absorption from the macropores into the dry or moist soil matrix. The process in which free water moves along macropores through an unsaturated soil matrix is called "bypass flow" (Bouma, 1984). Under ponding conditions (see Section V.A), macropores are completely filled with water and

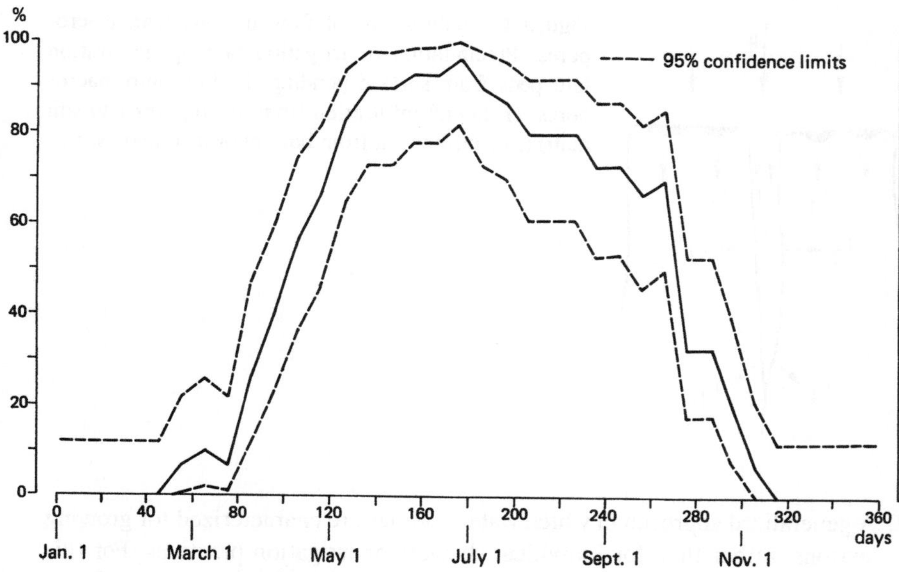

Figure 16. Probability (95% confidence limits) of adequate trafficability for every 10-day period calculated for a heavy clay soil. Probabilities were based on simulations for a 30-year period, using real weather condition and water-table levels (from Wösten and Bouma, 1985).

lateral absorption is relatively high. More common is a condition associated with sprinkling irrigation or rainfall in which macropores are only partly filled with water that runs vertically along their walls. Bouma and Dekker (1978) reported studies with dyes in the field indicating that only 1% or less of vertical ped faces in clay soil were covered by films of downward flowing water following sprinkling irrigation. Hoogmoed and Bouma (1980) proposed and tested a deterministic model defining vertical infiltration and lateral absorption of water in a cracked clay soil. They used morphological staining techniques to define the area available for lateral absorption in the cracks. Thus, morphological information is used in a unique manner in the flow model. Vertical infiltration patterns are bound to be different in different soil structures, but data on this are lacking. A deterministic model for ponded conditions was tested by Bouma and Wösten (1984). The basic approach being followed here is to define the complex flow system in the field into interacting subsystems (Figure 17). The magnitude of vertical infiltration into and lateral water movement away from macropores is determined by the number and continuity of those macropores. Soil survey descriptions of structure are helpful here, but these descriptions should perhaps be focused more sharply on defining pore patterns. Recently, comparable research results have been reported for solute transport (Hatano et al., 1985) and for water uptake by roots (Hasegawa and Sato, 1987).

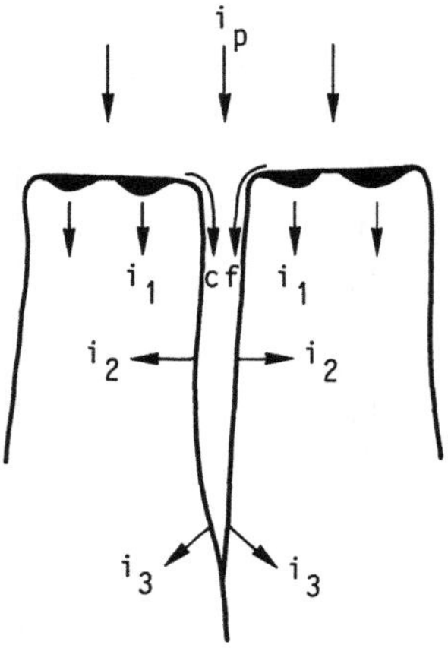

Figure 17. Subsystems of flow in soils with macropores. Precipitation or irrigation rate, i_p; infiltration into peds from surface ponding, i_1; flow into macropores, cf; lateral infiltration from macropores into soil matrix, i_2; infiltration from bottom of macropores, i_3.

2. A generalized approach in which water regimes are characterized for growing seasons, rather than for individual showers or irrigation practices. For this approach, bypass flow data are measured and are defined as a function of rainfall duration and intensity and initial water content of the soil. A simple field technique was devised to make many measurements in a relatively short period of time (Bouma et al., 1981). When rainfall or sprinkling intensity is known, bypass flow can be estimated. All water that is not part of bypass flow will infiltrate at the soil surface (i_1 in Figure 17) while water that moves into the macropores is absorbed laterally (i_2 in Figure 17) or at the depth where the macropores end (i_3 in Figure 17). In the heavy clay soil studied by Bouma and de Laat (1981), water moved into a subsoil with continuous cracks to the effect that the water was effectively removed to surrounding ditches.

The procedure discussed here consists, in fact, of reducing rainfall or sprinkling by a measured or estimated quantity of bypass flow to obtain the amount or estimated quantity of water infiltrating at the surface. Water taking part in bypass flow itself may partly be adsorbed laterally and it may accumulate at some depth. Bouma and de Laat (1981) reported 20% bypass flow in the summer months when considering natural rainfall patterns. Simulation results only agreed with field moisture data when bypass data were taken into account (e.g., Figure 4). Van Stiphout et al. (1987) demonstrated the effect of what they called internal catchment of water in a 60-cm-thick deposit of clay on top of a sandy loam sediment (Figure 18). A field infiltration experiment indicated wet conditions at 60 cm and 110 cm depth after two infiltration runs of 1 cm of water. Soil morphological descriptions, including staining of water-conducting macropores, indicated that cracks ended at 60 cm depth and that worm channels extended to a depth of 120 cm. A bypass flow routine was built into the traditional flow model, and was used successfully to predict the observed wetting pattern. Soil morphological descriptions were crucial to explain the results of the field measurements, which appeared to be highly erratic at first sight because the traditional flow model, which ignores macropores, predicted soil wetting to a depth of only 4 cm. Soil

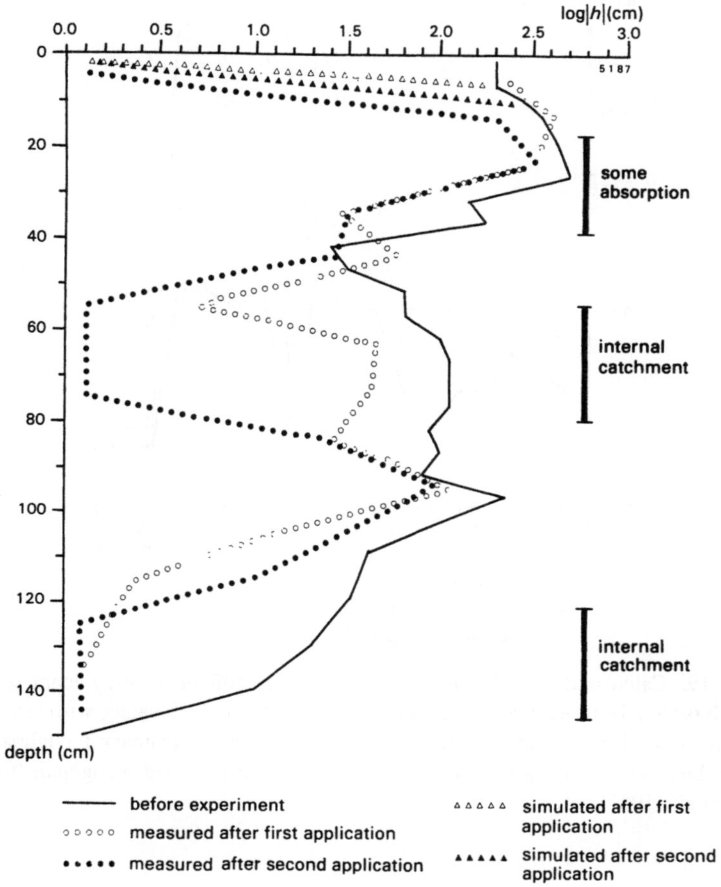

Figure 18. Water distribution after infiltration of two showers of 10 mm each in a 60-cm-thick clay soil with macropores overlying a sandy loam subsoil. Cracks extend to 60 cm and worm channels to 120 cm depth. Additions of water by bypass flow can be observed at both levels (from Stiphout et al., 1987).

morphological descriptions are also important to define the volume of water that is laterally absorbed from the macropores. Obviously, this volume is higher as the number of water-conducting macropores is higher.

The examples cited here relate to water movement. The practical relevance of processes described is particularly evident when dealing with solute movement or with wastewater (Dekker and Bouma, 1984; White et al., 1983; Leeds-Harrison et al., 1982).

D. Structural Phases of Soil Series

So far, reference has been made to legends of soil maps and to soil classifications, as if units being distinguished showed a unique characteristic behavior. This is,

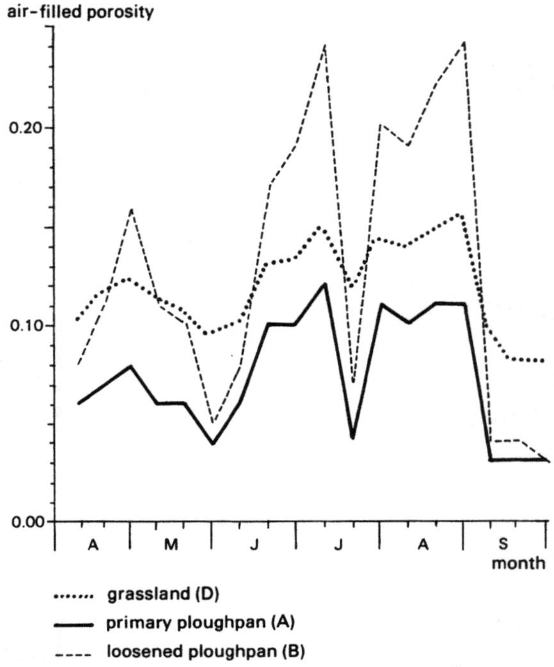

Figure 19. Calculated air-filled porosities in surface soil of a sandy loam soil for a 6-month period for three types of soil structures. Relatively high values with low fluctuations are found for grassland. Low values are found for the primary ploughpan while unfavorable, highly fluctuating values are found for the loosened ploughpan (from van Lanen et al., 1987).

of course, not necessarily true. Identical soils from a pedogenetic point of view may behave differently in terms of physical, chemical, or biological processes. For example, when a particular soil is being used as meadow, its structure is bound to be different from an identical soil that is being tilled with heavy machinery. Such different soil structures are, for example, associated with different hydraulic properties (Kooistra et al., 1984, 1985). Soil structure characterizations are useful for defining specific structure types that are associated with well defined types of soil management. Such characterizations should be supported by physical, chemical, and soil mechanical data, and distinction of specific structure types should be based on a statistical analysis of these supporting data to allow only major distinctions to be made.

This aspect is important, because soil morphological studies in the past have often suffered from a "classification syndrome," yielding highly detailed, descriptive schemes with a degree of detail that originated in the mind of the classifier rather than from a functional focus. Kooistra et al. (1984, 1985) defined a limited number of structure types for a Dutch sandy loam soil. These types had significantly different hydraulic conductivity and moisture retention characteristics, and were so different from a soil morphological point of view that different observers could arrive at identical classifications. Physical data obtained for the various soil structure types were used by van Lanen et al. (1987) in a dynamic simulation model for determining important land qualities such as moisture availability, aeration status, and trafficability. Considerable differences were found between grassland soil and tilled soil, as is illustrated in Figure 19 for the air-filled porosity as calculated for a real-time period of 6 months.

When characterizing soil types, soil structure differences should therefore be considered, particularly when considering detailed studies at large scales (e.g., 1:10,000 and larger). At random measurements within delineated areas of a soil map, which are frequently made in the context of (geo)statistical sampling procedures, may produce unnecessary variation when distinctly different structure types are present in one soil type. The apparent variability is reduced when all structure types are lumped together.

Results of van Lanen et al. (1987) demonstrate that land qualities are not only a reflection of soil physical properties, but also, of course, of environmental factors, such as climatic conditions. Even though hydraulic conductivity curves of the primary and secondary ploughpans described by Kooistra et al. (1984) were significantly different, those differences did not result in significantly different physical land qualities as determined by simulation over a 30-year period. This result reflects the character of the Dutch ocean climate.

The preceding discussion is meant to illustrate that soil structure descriptions are necessary to define what should be called "soil structure phases of well defined soil series." These structure types reflect certain types of soil management. As such, they function as indicators to a trained pedologist in the same manner that a clay cutan is an indicator of clay illuviation processes.

E. Spatial Interpolation

When discussing modern soil survey interpretations in Section III.B, the common procedure of defining a "representative" soil profile for each mapping unit was mentioned. Interpretations are based on the properties of such representative profiles, which are defined by soil surveyors. This procedure is rather subjective and relatively large differences may occur among surveyors (e.g., de Gruyter and Marsman, 1985). The "representative" profile does not indicate variation within the mapping unit, which is important because it determines the variability of soil behavior within the unit. Further, when making interpretations based on soil maps, the same abrupt boundaries that occur between soil units are found between areas assumed to exhibit a particular type of behavior (see Figures 6 and 8 in Section III.B). In reality, of course, these boundaries are not abrupt. One future possibility may be the application of "fuzzy" clustering techniques, which take into account that natural boundaries are usually gradual. Discussion of this topic is clearly beyond the scope of this chapter, as is a general discussion on spatial variability and interpolation. These topics have received

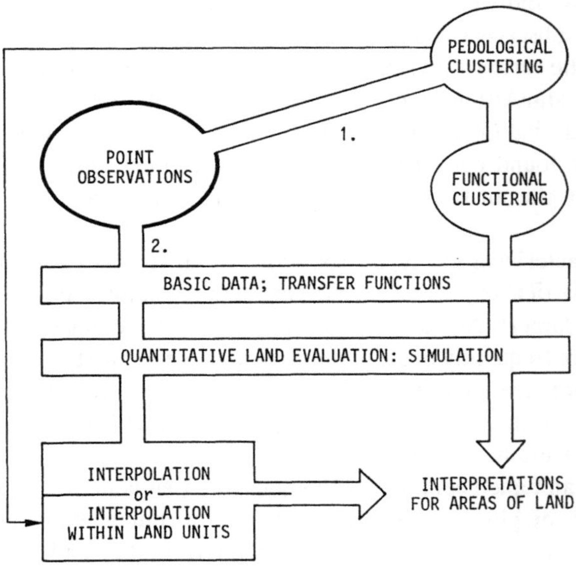

Figure 20. Diagram illustrating two procedures that result in soil survey interpretations for areas of land, using survey data (1) or point observations (2).

much recent emphasis, and excellent reviews have been presented by Webster (1985) and Burrough (1986).

The main item for discussion here is the observation that soil maps *are* an expression of spatial variability patterns, be it in a rather qualitative manner at a particular scale. Much work has been done to characterize the internal variability of mapping units in terms of soil properties (e.g., Wilding and Drees, 1983; de Gruyter and Marsman, 1985). However, little emphasis has been placed so far on comparing *interpretations* based on soil maps with those based on interpolation of point data. This would be a realistic exercise, because hydrologists and soil physicists are not necessarily convinced that boundaries on soil maps are relevant expressions to be used for sample stratification. They would rather make regular observations in a grid, to be followed by various types of interpolation (Nielsen and Bouma, 1985). However, Webster (1985) has remarked: "the sampling scheme was deliberately designed to distinguish major geological formations in the first stage, since it would have been foolish to have ignored such obvious features." The question remains as to what are "major formations" and "obvious features." Certainly not all boundaries on soil maps will qualify. Which boundaries do qualify?

A final observation must be made regarding point observations, which form the core of soil survey work. In earlier times borings by the soil surveyor were made only to check the mapping unit being encountered. No detailed record was made, let alone preserved. Currently, all separate borings are described in detail, and are recorded in Soil Information Systems. This allows classification and interpretation for each *point* observation.

Areal delineations can be derived by interpolation if so desired. The scheme of Figure 20 summarizes this discussion. Point observations are central. Pedological clustering results in a soil map (procedure 1); functional clustering was described in Section IV.B. Calculations then are made for "representative" units on the functional map derived from the soil map. The alternative procedure (no. 2) is to make calculations for each point observation. Data obtained can be interpolated without considering patterns of land units or by making interpolations separately within each land unit, on the basis of pedological clustering. Both procedures yield interpretations for areas of land. van Kuilenburg et al. (1982) and Bregt et al. (1987) compared estimates of the soil moisture supply capacity obtained by the two procedures and concluded that results were comparable. Comparisons of the two interpolation procedures, as shown in Figure 20, have not yet been made. Potentially, use of soil maps in this context would represent innovative use of existing data. The basic question being raised here is whether to "average first and then calculate" or "to calculate first and then to average" (de Wit and van Keulen, 1987). In any case, quantitative procedures require specific characterization of separate soil borings, in the context of future soil surveys.

V. Future Developments

Soil Survey and land evaluation have always been nourished by field observations and measurements in areas of land and by consideration of the economic and social context of any particulate problem. Modern technological developments have a major impact on future procedures. In fact, soil survey as we know it now is not likely to exist in another decade.

However, three important elements of soil survey and pedology are of permanent significance: (1) a focus on *areas* of land that occur in a geographical context; (2) emphasis on field observations and measurements in undisturbed field soils either to obtain data for simulation modeling or to validate simulation results; and (3) the need to integrate knowledge from a wide range of disciplines from geology to socioeconomics. Why claim a role here for soil survey specialists or pedologists while other specialists may be more proficient in applying physical, chemical, (geo)statistical, or socioeconomic techniques? As discussed, a pedologist is by training an integrator, and integration of specialized abilities will be very much needed in the future. Generalists with the ability to use modern integrating tools, such as systems analysis, have an important role to play in our future society. The following aspects are likely to play a role in soil science:

1. Remote sensing techniques will provide an almost continuous record of conditions at the soil surface in terms of such factors as roughness, temperature,

water content, and crop conditions. Pattern recognition techniques can be used to define subareas of land that show comparable behavior. These subareas may partly coincide with mapping units of soil survey but not necessarily so. Soil observations will, however, still be needed to explain observed differences. For example, an observed reduction of crop evapotranspiration can result from several different factors such as thickness and texture of topsoil, occurrence of coarse sandy layers in the subsoil that impede upward unsaturated flow, local soil compaction, or a locally deep water table. Each of these factors is associated with a different management procedure that is to be applied to improve moisture supply.

2. Automatic monitoring sensors will be developed and installed in situ, and they again will generate an almost continuous flow of data on soil water content, temperature, oxygen content, etc. Such data in combination with remote sensing data will allow testing and further development of accurate simulation models that can be used to predict *potential* conditions as a function of alternative land use scenarios.
3. Geographical information systems (GIS), to be fed by both vectorial and raster data, will allow instant production of interpretative maps. This will allow direct interaction with users of geographical information. The impact of this development can hardly be overestimated. Looking for data in files and drawing of maps usually requires many months. The formulation of problems to be studied tends therefore to have a rather static character, which is reflected in the type of answers being given. Interactive use of GIS systems allow a more dynamic exploration of what the user really wants, offering tailor-made solutions.
4. Computerized soil management systems are being developed that keep track of water and nutrient fluxes in the soil, allowing optimization of quantities of fertilizers to be applied and timing of application. By using dynamic simulation models and historic climatic data, it will be possible to predict probabilities of occurrence of soil conditions at a given date. This may refer to water contents of the soil, trafficability, aeration status, etc. (van Lanen et al., 1987). Such models can also be used in a predictive manner by feeding real-time data from short-range weather forecasts, which has the potential to allow the farmer to better plan his field work. A key element for running the simulation model is use of representative input data. Consideration of soil structure heterogeneity and spatial variability is important here, and pedologists are in a unique position to provide specific input.

Acknowledgments

The Dutch work being reviewed in this chapter was largely executed at the Netherlands Soil Survey Institute (STIBOKA), where the author was Deputy Director from 1983 to 1986. He gratefully acknowledges the cooperation of many STIBOKA colleagues.

References

Anderson, D.L., and J. Bouma. 1973. Relationships between hydraulic conductivity and morphometric data of an argillic horizon. *Soil Sci. Soc. Am. Proc.* 37:408–413.

Baker, F.G. 1978. Variability of hydraulic conductivity within and between nine Wisconsin soil series. *Water Resour. Res.* 14:103–108.

Baker F.G., and J. Bouma. 1975. Variability of hydraulic conductivity in two subsurface horizons of two silt loam soils. *Soil Sci. Soc. Am.* 40:219–222.

Belmans, C., Wesseling, J.G., and Feddes, R.A. 1983. Simulation model of the water balance of a cropped soil: SWATRE. *J. Hydrol.* 63:271–286.

Beven, K., and P. Germann. 1982. Macropores and water flow in soils. *Water Resour. Res.* 18:1311–1325.

Bie, S.W., and A. Ulph, 1972. The economic value of soil survey information. *J. Agric. Econ.* 23(3):285–297.

Bouma, J. 1981. Soil survey interpretation: estimating use-potentials of clay soil under various moisture regimes. *Geoderma* 26(3):165–177.

Bouma, J. 1982. Measuring the hydraulic conductivity of soil horizons with continuous macropores. *Soil Sci. Soc. Am. J.* 46(2):438–441.

Bouma, J. 1984. Using soil morphology to develop measurement methods and simulation techniques for water movement in heavy clay soils. In: J. Bouma and P.A.C. Raats (eds.), *Water and Solute Movement in Heavy Clay Soils.* Proc. ISSS Symp. ILRI, Publ. 37, pp. 298–316. Wageningen, the Netherlands.

Bouma, J. 1985. Graphical field estimation of the upward flux of water from the water-table to the rootzone in layered subsoils. *Soil Survey and Land Evaluation* 5(2): 34–39.

Bouma, J., 1986. Using soil survey information to characterize the soil water state. *J. Soil Sci.* 37(1):1–7.

Bouma, J., and L. W. Dekker. 1978. A case study on infiltration into dry clay soil. I. Morphological observations. *Geoderma* 20(1):27–40.

Bouma, J., and P.J.M. de Laat. 1981. Estimation of the moisture supply capacity of some swelling clay soils in the Netherlands. *J. Hydrol.* 49(3/4):247–259.

Bouma, J., and H.A.J. van Lanen. 1987. Transfer functions and threshold values: from soil characteristics to land qualities. In: *Quantified Land Evaluation*, Proceedings of a workshop. ISSS/SSSA, Washington, D.C. ITC Publ. Enschede, the Netherlands.

Bouma, J., and J.H.M. Wösten. 1984. Characterizing ponded infiltration in a dry cracked clay soil. *J. Hydrol.* 69:297–304.

Bouma, J., C.F.M. Belmans, and L.W. Dekker. 1982. Water infiltration and redistribution in a silt loam subsoil with vertical worm channels. *Soil Sci. Am. J.* 46(5):917–921.

Bouma, J., L.W. Dekker, and J.C.F.M. Haans. 1980. Measurement of depth to water table in a heavy clay soil. *Soil Sci.* 130(5):264–270.

Bouma, J., L.W. Dekkerk, and C.J. Muilwijk. 1981. A field method for measuring short-circuiting in clay soils. *J. Hydrol.* 52(3/4):347–354.

Bouma, J., P.J.M. de Laat, A.F. van Holst, and Th.J. van de Nes. 1980. Predicting the effects of changing water-table levels and associated soil moisture regimes for soil survey interpretations. *Soil Sci. Soc. Am. J.* 44(4):797–802.

Burrough, P.A. 1986. Principles of geographical information systems for land resources assessment. Oxford University Press, Oxford England.

Breeuwsma, A., J.H.M. Wösten, J.J. Vleeshouwer, A.M. van Slobbe, and J. Bouma 1986.

Derivation of land qualities to assess environmental problems from soil surveys. *Soil Sci. Soc. Am. J.* 50(1):186–190.

Bregt, A.K., J. Bouma, and M. Jellinek. 1987. Comparison of thematic maps derived from a soil map and from kriging of point data. *Geoderma* 39:281–291.

Burke, W., D. Gabriels, and J. Bouma (eds.) 1986. *Soil Structure Assessment*. Method manual sponsored by the Eur. Comm. Dir. Gen. VI (Agriculture). Balkema, Rotterdam/Boston.

de Gruyter, J.J., and B.A. Marsman. 1985. Transect sampling for reliable information on mapping units. In: D.R. Nielsen and J. Bouma (eds.), *Soil Spatial Variability*, pp. 150–166. Pudoc, Wageningen, Netherlands.

Dekker, L.W., and J. Bouma 1984. Nitrogen leaching during sprinkler irrigation of a dutch clay soil. *Agric. Water Manage.* 8(1):37–47.

de Laat, P.J.M. 1980. Model for unsaturated flow above a shallow water table, applied to regional sub-surface flow problem. *Agric. Res. Rep. 895*. Pudoc, Wageningen, Netherlands.

de Wit, C.T., and H. van Keulen. 1987. Modelling production of field crops and its requirements. *Geoderma* 40:253–267.

FAO. 1976. A framework for land evaluation. *Soils Bull. 32*. FAO, Rome, Italy.

FAO. 1983. Guidelines: land evaluation for rainfed agriculture. *Soils Bull. 52*. FAO, Rome, Italy.

FAO-UNESCO. 1987. *Soil Map of the World* (revised legend; amended fourth draft). World Resources Report 60. FAO, Rome, Italy.

Feddes, R.A., P.J. Kowalik, and H. Zaradny. 1978. *Simulation of Field Water Use and Crop Yield*. Simulation monographs, Pudoc, Wageningen, Netherlands.

Green, R.E., and J.C. Corey. 1971. Calculation of hydraulic conductivity: A further evaluation of some predictive methods. *Soil Sci. Soc. Am. Proc.* 35:3–8.

Haans, J.C.F.M. (ed.) 1979. Interpretation of soil maps. *Stiboka Report 1463* (in Dutch). Netherlands Soil Survey Institute, Netherlands.

Hasegawa, S., and T. Sato. 1987. Water uptake by roots in cracks and water movement in clayey subsoil. *Soil Sci.* 143(5):381–386.

Hatano, R., T. Sakuma, and H. Okagima. 1985. The source-sink effect of clayey soil peds on solute transport. *Soil Sci. Plant Nutr.* 31(2):199–213.

Hoogmoed, W.B., and J. Bouma. 1980. A simulation model for predicting infiltration into cracked clay soil. *Soil Sci. Soc. Am. J.* 44(3):458–461.

Jury, W.A. 1982. Simulation of solute transport using a transfer functions model. *Water Resour. Res.* 18:363–368.

Jury, W.A., L.H. Stolzy, and P. Shense. 1982. A field test of the transfer function model for predicting solute transport. *Water Resour. Res.* 18:369–375.

Klute, A. 1973. Soil Water Flow Theory and its application in field situations. In: R.R. Bruce, K.W. Flach, and H.M. Taylor (eds.), *Field Soil Water Regime*, Soil Sci. Soc. Am. Special Pub. No. 5, pp. 9–31. Soil Science Society of America, Madison, Wisconsin.

Kooistra, M.J., J. Bouma, O.H. Boersma, and A. Jager. 1984. Physical and morphological characterization of undisturbed and disturbed ploughpans in a sandy loam soil. *Soil & Tillage Res.* 4:405–417.

Kooistra, M.J., J. Bouma, O.H. Boersma, and A. Jager. 1985. Soil structure variation and associated physical properties of some Dutch Typic Haplaquents with sandy loam texture. *Geoderma* 36:215–229.

Lamp, J., and W. Kneib. 1981. Zur quantitativen Erfassung und Bewertung von Pedofunktionen. *Mitt. Dtsch. Bodenkundl. Gesellschaft* 32:695–711.

Lauren, J.G., R.J. Wagenet, J. Bouma, and J.H.M. Wösten, 1987. Variability of saturated hydraulic conductivity in a Glossaquic Hapludalf with macropores. *Soil Sci.* 145:20–28.

Leeds-Harrison, P., G. Spoor, and R.J. Godwin. 1982. Water flow to mole drains. *J. Agric. Eng. Res.* 27:81–91.

McKeague, J.A., R.G. Eilers, A.J. Thomasson, M.J. Reeve, J. Bouma, R.B. Grossman, J.C. Favrot, M. Renger, and O. Strebel.1984. Tentative assessment of soil survey approaches to the characterization and interpretation of air-water properties of soils. *Geoderma* 34:69–100.

Nielsen, D.R., and J. Bouma (eds.). 1985. *Soil Spatial Variability.* Proceedings of an ISSS-SSSA workshop. Pudoc, Wageningen, Netherlands.

van Genuchten, M.Th. 1980. A closed-formed equation for predicting the hydraulic conductivity of unsaturated soils. *Soil Sci. Soc. Am J.* 44:892–989.

van Keulen, H., and J. Wolff (eds.) 1986. Modelling of agricultural production: weather, soils, crops. Pudoc, Wageningen, Netherlands.

van Kuilenburg, J., J.J. de Gruyter, B.A. Marsman, and J. Bouma. 1982. Accuracy of spatial interpolation between point data on soil moisture supply capacity, compared with estimates from mapping units. *Geoderma* 27(4):311–325.

van Lanen, H.A,J., M.H. Bannink, and J. Bouma. 1987. Use of simulation to assess the effect of different tillage practices on land qualities of sandy loam soil. *Soil & Tillage Research* 10:347–361.

van der Sluijs, P., and J.J. de Gruijter. 1985. Water table classes: a method used to indicate seasonal fluctuation and duration of water tables on Dutch soil maps. *Agric. Water Manage.* 10(2):109–125.

Soil Survey Staff (1975). *Soil Taxonomy. A Basic System of Soil Classification for Making and Interpreting Soil Surveys.* U.S.D.A. Agric. Handbk 436. U.S. Dept. of Agriculture, Washington, D.C.

van Stiphout, T.P.J., H.A.J. van Lanen, O.H. Boersma, and J. Bouma. 1987. The effect of bypass flow and internal catchment of rain on the water regime in a clay loam grassland soil. *J. Hydrol.* 95:1–11.

Wagenet, R.L., and J.L. Hutson. 1987. *LEACHM Version 1.0.* Dept. of Agronomy, Cornell University, Ithaca, New York.

Webster, R. 1985. Quantitative spatial analysis of soil in the field. In: B.A Stewart (ed.), *Advances in Soil Science*, Vol. 3, pp. 1–71. Springer-Verlag, New York.

White, R.E. 1985. The influence of macropores on the transport of dissolved and suspended matter through soil. In: B.A. Stewart (ed.), *Advances in Soil Science*, Vol. 3, pp. 95–121. Springer-Verlag, New York.

White, R.E., S.R. Wellings, and J.P. Bell. 1983. Seasonal variations in nitrate leaching in structured clay soils under mixed land use. *Agric. Water Manage.* 7:391–410.

Wilding, L.P., and L.R. Drees, 1983. Spatial variability and pedology. In: L.P. Wilding, N.E. Smeck, and G.F. Hall (eds.), *Pedogenesis and Soil Taxonomy. I. Concepts and Interactions.* Developments in Soil Science, 11A, pp. 83–113. Elsevier, Amsterdam.

Wösten, J.H.M., and J. Bouma. 1985. Using simulation to define moisture availability and trafficabililty for a heavy clay soil. *Geoderma* 35(3):187–197.

Wösten, J.H.M., M.H. Bannink, and J. Bouma. 1987. Land evaluation at different scales: you pay for what you get! *Soil Survey and Land Evaluation* 7:13–24.

Wösten, J.H.M., J. Bouma, and G.H. Stoffelsen. 1985. The use of soil survey data for regional soil water simulation models. *Soil Sci. Soc. Am. J.* 49(5):1238–1245.

Index

A

Africa, 2, 59
Agricultural wastes, 130
Alfalfa, 59, 62, 66, 70, 73, 75, 76
Alfisols, 53
Allium cepa, 133
Altavista soil, 95
Aluminum, 4, 55–58, 71, 75, 78, 79, 81, 83, 84, 86, 89–95, 101, 136, 147–149, 155, 158, 161, 163, 164, 166, 199
Ammonium, 86
Ammonium sulfate, 96
Antimony, 116
Apple, 59, 64, 68, 73, 74, 97
Appling soil, 57, 71, 73, 78, 82, 83, 86–88
Arid regions, 8, 15
Arizona, 3
Arsenic, 4, 98, 116, 118
Atriples spp., 123, 124
Australia, 8, 9, 53, 54, 97, 118, 119, 121, 127, 132, 136
Available water, 182, 186
Avocados, 97

B

Balmoral soil, 56
Basaluminite, 92
Beans, 9, 64, 66, 73, 129, 135
Benjamin Franklin, 96
Bentonite, 13, 160
Birnessite, 166
Bismuth, 116
Boehmite, 161
Boron, 115
Botucatu soil, 66
Brachiaria decumbens, 66
Brazil, 55–59, 64, 66, 67, 72, 77, 78, 91, 94, 101
Bulk density, 16, 39, 52, 73, 196–198, 202

C

C_4 plants, 115
Cabbage, 129
Cadmium, 4, 98, 99, 114, 115, 117–120, 123–125, 128, 129, 133, 135, 136, 149, 152, 155, 156, 166, 167
Calcium, 2–8, 10–12, 17, 19, 20, 55–58, 67–71, 77–79, 81, 83, 84, 86, 87, 89–91, 94, 97, 100, 101, 146, 149, 157, 168
California, 18, 50
Canada, 3, 118, 121
Carrot, 129
Cation-exchange capacity, 13, 45, 51, 90, 95, 137, 153, 199, 201
Cattle, 130
Cecil soil, 19, 27, 45
Central America, 2
Chloride, 81
Chlorine, 13, 20, 21, 83
Chromium, 116, 125, 153
Cloride, 82
Coal, 115
Cobalt, 115, 116, 147, 151, 152, 156, 159, 167

Coffee, 59, 64, 68
Compaction, 25, 26, 36, 39, 48, 52, 210
Cone index, 87
Copper, 4, 69, 70, 113–119, 123, 124, 128, 129, 134, 135, 146, 150, 152, 155–157, 159, 161, 164, 166, 169, 170
Corn, *see also* Maize, 64, 68, 76, 98, 99
Cotton, 53, 59, 62, 68
Crop residues, 53
Crust, 23, 24, 26, 27, 34, 36, 37, 39, 41, 44, 47, 50, 52
Crusting, 2, 6, 19, 28, 48, 49, 53, 54, 100
Cultivation, 7, 8, 39

D
DRIS, 71
Darcian principles, 201
Davidson soil, 19, 95
Davis equation, 22
Dc arc spectrophotometer, 168
Dietary limits, 137
Diffuse double layer, 7, 10, 12, 14
Double-ring infiltrometer, 39
Drainage, 179–183
Dryland farming, 39, 59

E
Earthworms, 128, 192, 193
Electric conductivity, 5
England, 3
Erosion, 2, 6, 9, 39, 43, 47, 48
Europe, 3, 96, 118
Evapotranspiration, 182, 196
Exchangeable sodium percentage, 8, 9, 11, 12, 15, 17, 18, 19, 21, 24–26, 28, 33–37, 43, 50

F
Fallbrook soil, 17, 18
Feldspars, 17
Fertilization, 185
Field capacity, 83, 181
Florida, 3, 4, 54, 97, 98
Fluorine, 4, 83, 86, 89, 90, 116
Food Security Act, 54
Forage, 68
France, 3
Fulvic complexes, 160

G
Gapon equation, 22
Georgia, 19, 27, 44, 45, 53, 54, 57, 62, 66, 73, 77, 78, 87
Geothite, 148
Gibbsite, 18, 91, 92, 161, 164, 166
Gila soil, 18
Glycine max., 118
Goethite, 160–163, 166, 167
Golan soil, 20, 21
Grassed waterways, 98
Greenland, 126
Greenville soil, 27, 28, 53
Gypsum, 2–101

H
Haematite, 160, 166
Halloysite, 18, 92
Hamra-Netanya soil, 26
Haploxeralf soil, 20, 21
Hematite, 167
Herbicides, 185
Hornblende, 17
Humic complexes, 160
Humid climates, 44
Humid regions, 6, 9, 27, 101
Humid tropics, 100
Hydraulic conductivity, 7–9, 13, 15–18, 20, 21, 28, 39, 50, 182, 184, 192, 193, 196–198, 206, 207

I
Illite, 8, 12, 19, 50
Inceptisol, 64
India, 52
Infiltration, 2, 6, 9, 15, 20, 22–29, 31–34, 36, 44, 46, 47, 50, 52, 53, 88, 100, 128, 192–195, 202–205
Iodine, 116
Iowa, 3
Iredell soil, 19

Iron, 3, 4, 18, 68, 115, 116, 146, 147, 149, 152, 153, 157–159, 161–163, 170, 178, 199
Irrigation, 15, 19, 24, 42, 50, 117, 120, 136, 194, 203
Israel, 5,27, 39, 44, 45

J

Jurbanite, 91, 92

K

Kaolinite, 8, 13, 14, 18, 19, 27, 28, 44, 89, 91, 92

L

Lactuca sativa, 133
Land evaluation, 177–210
Land quality, 178, 179, 186, 196, 197, 200, 207
Latossdo soil, 57
Lead, 4, 114, 115, 118, 119, 122, 123, 125–130, 149, 150, 155, 156, 166, 167, 169
Leek, 129
Legumes, 2
Lepidocrocite, 163
Lettuce, 129, 133, 134
Lewis acids, 151
Lime, 34, 54–56, 88, 97
Lithiophorite, 162, 166
Lolium perenne, 134

M

Macropores, 184, 193, 194, 201–205
Maghemite, 163
Magnesium, 3, 4, 17, 22, 56, 57, 67, 69, 70, 77–79, 86, 94–96, 146, 149, 168
Magnetite, 160, 163, 167
Maize, *see also* Corn, 55, 58–60, 72–74, 134
Manganese, 58, 68, 69–71, 83, 86, 115, 116, 146–148, 152, 153, 157–159, 161, 162, 164, 165
Matric suction gradient, 23
Mercury, 114
Mica, 18
Microorganisms, 128
Miscible displacement, 22
Modulus of rupture, 20
Moisture availability, 186
Moisture supply capacity, 179–181
Molybdenum, 115, 116
Montmorillonite, 10, 11, 13, 14, 16, 18, 27, 161
Mossbauer techniques, 148
Mulch, 7, 34

N

N_2 Fixation, 128
Nahal-Oz soil, 16, 20, 21, 25, 31
Natroalunite, 92
Netanya soil, 16, 18, 24, 27, 36–38
Netherlands, 118, 179, 187, 188
New Mexico, 3
New York, 3
New Zealand, 97, 132
Nickel, 114, 116, 119, 125, 128, 129, 134, 150, 156, 166, 167
Nitrate, 81, 83, 86
Nitrogen, 55, 68–70, 72, 128, 181
No-till, *see also* Zero tillage, 53, 54, 100, 101
Nontronite, 162
Normandien soil, 57, 79, 83, 86
North Carolina, 13
Nova Scotia, 3, 4

O

Onion, 133, 134
Organic carbon, 57, 159
Organic matter, 8, 44, 72, 128, 136, 144, 146–148, 150–159, 161, 162, 164–166, 169, 181, 196–199, 202
Organic wastes, 144
Oxisols, 54, 56, 57, 84, 90, 91

P

Pachappa soil, 18
Pasture crops, 54
Pasture plants, 130

Pea, 129
Peaches, 59, 62
Peanuts, 68
Peat, 114
Pedology, 178
Penetration resistance, 20, 87
Percolation, 5, 14, 16, 17, 24
Permeability, 52
Pesticides, 185
pH, 5, 7, 8, 13, 14, 19, 45, 56, 75, 76, 78, 79, 81–84, 86, 90, 93, 117, 120, 127, 131, 135–137, 144–165, 171
Phaseolus vulgaris, 135
Phosphate, 14, 54, 82, 86, 94, 98, 118, 147, 148, 164
Phosphate sorption capacity, 199, 200, 201
Phosphogypsum, 2–101
Phosphoric acid, 3
Phosphorus, 4, 68–70, 72, 89, 128
Phytophthora cinnamoni, 97
Piezometers, 193
Pig manure, 170
Pig slurries, 119
Plaggept soil, 199, 201
Plagioclase, 17
Plinthic Paleudult, 57
Ploughpan, 206
Plow, 32
Porosities, 206
Potassium, 4, 57, 67–70, 77–79, 83, 86, 95, 96, 147
Potassium nitrate, 92
Potato, 129
Poultry waste, 119

R

Radioactivity, 97–99
Raindrop impact, 7, 8, 15, 23, 25–28, 34, 36, 39, 41, 43, 44, 47, 48
Raindrop splash, 44
Rainfall, 182, 194, 204
Rainfall distribution, 2
Rainfall simulator, 24, 26, 27, 32, 39, 46, 49, 52
Redox, 127, 136
Reduced tillage, 19
Remote sensing, 209
Rhizosphere, 75, 76
Rhodic Paleudult, 28
Rhodoxeralf soil, 20, 27, 45, 48
Rice, 59, 64, 73
Rock phosphate, 3, 117
Roman Empire, 114
Runoff, 2, 5, 6, 24, 27, 28, 36, 39, 41–47
Ryegrass, 134

S

Sahel region, 2
Saline soils, 5
Saltbrush, 123, 124
Scanning electron microscope, 34–38, 144
Sediment basins, 98
Sediments, 120, 167
Selenium, 4, 98, 116
Semiarid climates, 44
Semiarid regions, 2, 4, 6, 8, 9, 15, 27, 39, 100
Sericia, 62
Sheep, 119, 123–125, 130
Silage, 61
Silica, 147, 166
Silicon, 4, 83, 86, 89
Simulation models, 184, 186, 195, 209, 210
Sludge, 119, 120, 133–136, 144–171
Smectite, 8, 13, 14, 16, 19, 44, 50
Sodic soils, 2, 8, 9, 16–20, 22, 27–29, 43, 53
Sodium, 4, 5, 7–15, 17–19, 21, 22, 24, 25, 42, 44, 52, 86, 115, 147
Sodium adsorption ratio, 9, 16, 17, 22
Sodium nitrate, 27, 28
Soil acidity, 2, 54–56, 58, 59, 81, 83, 84, 94, 100
Soil aggregates, 6, 7, 19, 25, 26, 28, 36, 39, 44, 45, 48, 50, 52, 54, 88, 89, 192
Soil amendments, 3, 116
Soil chemical properties, 2
Soil classification, 177
Soil conditioner, 2
Soil conservation, 43

Soil extractants, 144–171
Soil fauna, 128
Soil fertility, 43, 128, 181
Soil genesis, 177
Soil loss, 27
Soil physical properties, 2, 15, 66, 87, 100, 198, 199, 207
Soil pollution, 114–117, 120, 121, 123–126, 128–133, 136, 137, 144
Soil porosity, 23, 34
Soil potentials, 182
Soil productivity, 118
Soil resistance, 48
Soil roughness, 209
Soil sealing, 20, 25, 28, 34, 39, 45, 53
Soil sodicity, *see also* Sodic soils, 24–26, 35, 36
Soil strength, 19
Soil structure, 6, 7, 23, 40, 44, 101, 128, 157, 178, 195–197, 202, 203, 206, 207, 210
Soil survey, 177–210
Soil tilth, 19
South Africa, 56–60, 66, 72, 77–79, 92, 94, 97
South America, 2
Soviet Union, 3
Soybean, 59, 60, 66, 76, 98, 99, 118
Spatial variability, 36, 178, 185, 186, 208, 210
Spinach, 129
Sprinkler infiltrometer, 39
Sulfur, 163
Sulfur nutrition, 61, 63, 67–71, 96, 97, 100
Superphosphate, 97
Sweden, 118

T
Tasmania, 116
Terraces, 98
Texas, 3
Tifton soil, 83, 86, 95, 96
Tillage, 19, 53
Tin, 116
Titanium, 129
Tomato, 129
Trafficability, 179–183, 200, 202, 203, 210
Triticum vulgare, 118
Turnip, 129
Typic Hapludult, 57

U
Ultisols, 27, 54, 60, 62, 66, 73, 84, 87, 94, 100
United Kingdom, 118
United States, 2–4, 8, 9, 19, 27, 44, 45, 53, 54, 56, 58, 59, 72, 118, 132, 137
Universal Soil Loss Equation, 46

V
van der Waals forces, 7
Vanadium, 115, 116
Vermiculite, 8, 14, 19

W
Wastes, 146
Water conservation, 22
Water erosion, 44
Water table, 181, 193, 194, 197, 202
Wedowee soil, 45
Weed control, 19
Wheat, 39, 41, 53, 59, 64, 97–99, 118, 122
Wilting point, 181
Worsham soil, 45
Wyman soil, 51
Wyo soil, 51, 52

X
X-ray diffraction, 144, 168

Z
Zea mays, 134
Zero tillage, *see also* No-till, 9
Zinc, 68, 69, 114–116, 118, 119, 122, 125, 128, 129, 133–135, 149–153, 155–157, 161, 162, 166, 167

MIX
Papier aus verantwortungsvollen Quellen
Paper from responsible sources
FSC® C105338

If you have any concerns about our products,
you can contact us on
ProductSafety@springernature.com

In case Publisher is established outside the EU,
the EU authorized representative is:
Springer Nature Customer Service Center GmbH
Europaplatz 3, 69115 Heidelberg, Germany

Printed by Libri Plureos GmbH
in Hamburg, Germany